PROTEIN BIOCHEMISTRY, SYNTHESIS, STRUCTURE AND CELLULAR FUNCTIONS

# NEW RESEARCH ON MOLECULAR CHAPERONES

# Protein Biochemistry, Synthesis, Structure and Cellular Functions

Additional books in this series can be found on Nova's website under the Series tab.

Additional e-books in this series can be found on Nova's website under the e-book tab.

PROTEIN BIOCHEMISTRY, SYNTHESIS, STRUCTURE AND CELLULAR FUNCTIONS

# NEW RESEARCH ON MOLECULAR CHAPERONES

JAIME WYATT
EDITOR

nova publishers
*New York*

## NOTICE TO THE READER



Additional color graphics may be available in the e-book version of this book.

**Library of Congress Cataloging-in-Publication Data**

Library of Congress Control Number: 2015946968

ISBN: 978-1-63483-543-5

*Published by Nova Science Publishers, Inc. † New York*

# CONTENTS

# PREFACE

In this book, the current advances on the functional role of molecular chaperones in protozoan's mechanisms of response to stressful conditions, as well as an overview of studies focusing on molecular chaperones as potential targets for designing novel anti-parasitic drugs, are examined. In the second chapter, the current progress regarding the structure-function relationship as well as its association with the chaperone-like activity of the mammalian PDIA1/P4HB protein are discussed. The final chapter focuses on the cytoplasmic and nuclear functions of URI/Bud27 in different organisms.

Chapter 1 – *In vivo*, proteins are synthesized as linear chains by ribosomes. To be functional, must fold and get their three-dimensional structure (native conformation). Interestingly, some proteins can get the native conformation spontaneously; however, many others cannot and are likely to get a misfolded conformation. Therefore, to facilitate the folding process, the endoplasmic reticulum contains a molecular machinery that includes a set of chaperones, which prevents protein aggregation and promotes correct folding, thereby controlling cellular proteostasis. Despite this, some proteins fail to adopt a native conformation and are more likely to form aggregates, which represent a constant threat to cell function and viability. For instance, human amyloid diseases are characterized by the accumulation of unfolded proteins which form insoluble aggregates. Parasitic infections are a major public health issue in tropical and subtropical regions. Several studies focusing on the biology of medically-important protozoan parasites, such as Leishmania, Trypanosoma, and Plasmodium, have revealed processes related to their particular lifestyle. As they endure harsh environment conditions during its life cycle (e.g., dealing with at least four types of cellular stress: thermal, nutritional, osmotic, and oxidative), protozoan parasites have developed a response mechanism, which includes activation of molecular chaperones, to avoid unfolding of proteins. Consequently, molecular chaperones play a key role in both maintaining protein homeostasis and promoting correct folding of virulence-related proteins. This chapter reviews the current advances on the functional role of molecular chaperones and foldases from medically-important protozoan parasites in their mechanisms of response to stressful conditions, as well as an overview of studies focusing on these key molecules as potential targets for designing novel anti-parasitic drugs.

Chapter 2 – Protein folding is a sensitive process that can be affected by several factors, among which pH and temperature are notable (environmental factors). Protein misfolding is an important issue in some human disorders, e.g., Alzheimer's disease. Molecular chaperones are multi-domain proteins with a specific function in folding, unfolding, and remodeling of proteins. *In vivo*, molecular chaperones assist nascent polypeptides throughout the folding process until the native conformation has been attained. Furthermore, they can regulate unfolding and aggregation, protecting the cell from a proteotoxic stress. Protein disulfide isomerase (PDI, EC 5.3.4.1) enzymes are oxidoreductases that catalyze the oxidation and rearrangement of disulfide bonds in polypeptide substrates. PDI enzymes share as a common feature an active domain that folds similar to the bacterial protein thioredoxin (thus named thioredoxin domain), which has the conserved CXXC motif at its active site. The protein disulfide isomerase A1 (PDIA1), also named beta subunit of prolyl-4-hydroxylase (P4HB), is a multifunctional enzyme that plays a key role in the dithiol-disulfide equilibrium of the secretory pathway. Interestingly, a chaperone-like activity has been observed at high concentration, inhibiting aggregation of misfolded proteins. Conversely, anti-chaperone activity has been noted at low concentrations, promoting aggregation. This chapter reviews the current progress regarding the structure-function relationship as well as its association with the chaperone-like activity of the human PDIA1/P4HB protein.

Chapter 3 – The unconventional prefoldin Rpb5 interactor (URI/RMP), and its ortholog in yeast, Bud27, are members of the prefoldin (PFD) family of ATP-independent molecular chaperones. URI was originally identified as a protein binding the Rpb5 subunit of all three nuclear RNA polymerases (RNA pol) and is considered to function as a scaffold protein able to assemble additional members of the PFD family in both human and yeast.

URI and Bud27 have been linked to essential cellular processes both in the nucleus and the cytoplasm, such as RNA polymerases assembly, translation initiation, genome stability, DNA damage response, and viability. Furthermore, URI has been shown to be involved in several types of cancer processes and to regulate the TOR pathway, not only in humans, but also in yeast.

Taken together, this data points to a role for URI/Bud27 as a global molecular chaperone participating in multiple essential cellular processes.

In this review, the authors focus on the cytoplasmic and nuclear functions of URI/Bud27 in different organisms.

In: New Research on Molecular Chaperones
Editor: Jaime Wyatt
ISBN: 978-1-63483-543-5


*Chapter 1*

# PROTEIN FOLDING AND MOLECULAR CHAPERONES OF PROTOZOA

***Patricia L. A. Muñoz, Alexis Z. Minchaca, Rosa E. Mares and Marco A. Ramos****
Chemical Sciences and Engineering Faculty,
Autonomous University of Baja California, at Tijuana, Mexico

## ABSTRACT

*In vivo*, proteins are synthesized as linear chains by ribosomes. To be functional, must fold and get their three-dimensional structure (native conformation). Interestingly, some proteins can get the native conformation spontaneously; however, many others cannot and are likely to get a misfolded conformation. Therefore, to facilitate the folding process, the endoplasmic reticulum contains a molecular machinery that includes a set of chaperones, which prevents protein aggregation and promotes correct folding, thereby controlling cellular proteostasis. Despite this, some proteins fail to adopt a native conformation and are more likely to form aggregates, which represent a constant threat to cell function and viability. For instance, human amyloid diseases are characterized by the accumulation of unfolded proteins which form insoluble aggregates. Parasitic infections are a major public health issue

* To whom should correspondence be addressed: Marco A. Ramos, PhD., E-mail: mramos@uabc.edu.mx

in tropical and subtropical regions. Several studies focusing on the biology of medically-important protozoan parasites, such as Leishmania, Trypanosoma, and Plasmodium, have revealed processes related to their particular lifestyle. As they endure harsh environment conditions during its life cycle (e.g., dealing with at least four types of cellular stress: thermal, nutritional, osmotic, and oxidative), protozoan parasites have developed a response mechanism, which includes activation of molecular chaperones, to avoid unfolding of proteins. Consequently, molecular chaperones play a key role in both maintaining protein homeostasis and promoting correct folding of virulence-related proteins. This chapter reviews the current advances on the functional role of molecular chaperones and foldases from medically-important protozoan parasites in their mechanisms of response to stressful conditions, as well as an overview of studies focusing on these key molecules as potential targets for designing novel anti-parasitic drugs.

# 1. Introduction

Proteostasis, or protein homeostasis, is mainly referred to the cellular processes involved in regulation of the equilibrium between the synthesis, folding, unfolding, and degradation of proteins. Interestingly, molecular chaperones and foldases (folding enzymes) play key roles in those processes. *In vivo*, protein folding is driven by the concerted action chaperones and foldases, as they assist nascent polypeptides to adopt their correct three-dimensional structure. This task is not easy, since there are a wide range of possible conformations that can be adopted, especially in a conglomerated intracellular environment, affecting the selection of the correct protein folding path. So, to circumvent any detour, chaperones and foldases act cooperatively to remodel and reactivate their polypeptide substrates.

In cells, as all processes occur simultaneously, thigh regulation mechanisms must be active to preserve protein homeostasis. An imbalance in these processes could lead to accumulation of aberrant isoforms (unfolded and misfolded polypeptides), which promote the formation of aggregates, resulting in loss of function, cellular stress and, ultimately, cell death. Chaperones and foldases respond to these stressful conditions, assisting protein folding and, thus, preventing any further outcome. So far, the most studied are two heat shock proteins (Hsp70 and Hsp90), an oxidoreductase (protein disulfide isomerase, PDI), and a lectin (calreticulin, CRT).

Chaperones and foldases have been identified in all organisms studied to date, from prokaryotic to eukaryotic cells, including some human protozoan parasites (Plasmodium, Leishmania, Trypanosoma, Toxoplasma, Giardia, and Entamoeba). As expected, these organisms also depend on protein folding, assisted by chaperones and foldases, to fulfill their cellular functions. Protozoan parasites advance through different stages to accomplish its life cycle. Interestingly, the stage conversion includes several cellular adaptations intended to withstand adverse environmental conditions, from starvation and nutrient depletion to evasion of the host immune system.

Chaperones and foldases play an important role in the development and stage transition of protozoan parasites, promoting the correct folding of proteins involved in cell development, allowing resistance to unpleasant conditions and, thus, ensuring the success of the infection. Furthermore, for some intracellular parasites, such as Plasmodium, chaperones have been also associated with pathogenesis, suggesting a possible selection and expansion of its functions. Therefore, chaperones and foldases represent potential targets for the development of new or improved anti-parasitic drugs.

For instance, the antibiotic geldanamycin binds to the ATP binding site of the Hsp90 counterpart from several parasites and, as a consequence, ATP hydrolysis is inhibited. Since this activity is essential for assisting correct folding, some vital functions are interrupted, such as cell growth and proliferation. Furthermore, geldanamycin inhibits more efficiently chaperones from parasites than those from human. Albeit it has potential as an anti-parasitic agent, geldanamycin cannot be regarded as suitable for human therapy because it exhibits hepatotoxicity. Nonetheless, this compound has served as core to obtain derivatives with reduced toxicity, but keeping its selectivity.

In addition to Hsp90, other chaperones and foldases, such as Hsp40, Hsp70, CRT, PPI, and PDI, have potential as targets for development of anti-parasitic drugs. However, more studies leading to the discovery of new or improved chemotherapeutics are required; specially, targeting medically-important parasites.

## 2. Molecular Chaperone Machinery and Protein Folding

Cellular proteostasis, as a quality control mechanism, includes the regulation of protein synthesis, folding, unfolding, and degradation. Interestingly, chaperones and foldases are key players in maintaining this cellular equilibrium (Saibil 2013), since they participate in protein folding, a main process that is coupled to protein synthesis (Kim 2015).

During protein synthesis, the polypeptide folds into a native conformation (tertiary structure) as it emerges from the ribosome; however, many small proteins fail in their attempt to adopt a correct folding and remain as denatured polypeptide (Kim 2015). Two possible mechanisms have been proposed to describe the achievement of the native structure. The first comprises the formation of an initial core structure, from whom the remaining polypeptide chain adopts its correct folding. While in the second, regions of the polypeptide fluctuate in the formation of a secondary structure; then, an intermediate will be formed by joining altogether, which eventually develop the native structure (Efimov 2014).

In contrast, by using mechanical strength, chemical denaturation, and point mutations, it has been demonstrated that acquisition of the protein native structure involves a complex energy landscape in a composite cellular environment, where proteins are exposed to many disturbances that may strongly influence the selection of the folding path (Guinn 2015). Moreover, it has also been shown that during translation, the rate at which each amino acid is incorporated into the nascent polypeptide affects the probability of a protein to acquire its native conformation; therefore, protein folding during translation is hardly balanced (Caniparoli 2015).

Furthermore, to overcome difficulties in the protein folding pathway and to prevent aggregation, chaperones and foldases cooperate and regulate each other to reshape and reactivate substrate polypeptides. In cells, the existing combinations of chaperones and foldases determine the efficiency and specificity of the molecular folding machinery (Reidy 2014).

## 2.1. Heat Shock Proteins

The best preserved mechanism to protect cells from misfolded proteins is the refolding mediated by evolutionarily conserved heat shock proteins (Hsp). Hsp are induced in response to several types of stress, such as heat shock, nutrient deprivation, or mechanical damage. Hsp recognize unfolded or misfolded proteins and facilitate their refolding either by a mechanism dependent or independent of ATP (Rosenberg 2015; Brüning 2015).

**Small Heat Shock Proteins (sHsp)**

The sHsp family of chaperones is characterized by their low molecular weight (12-43 kDa) and by containing a conserved α-crystalline domain. This chaperone family is found in a wide variety of cell types, from prokaryotes to eukaryotes, with a number of members ranging from 1 to 19. Interestingly, sHsp assist protein homeostasis by preventing or reversing aggregation under conditions of proteotoxic stress. This function is carried out through an ATP-independent activity, as holdases, by binding partially unfolded polypeptide to an assembly of sHsp oligomers. Then, when conditions are optimal, polypeptide substrates are released for refolding with the assistance of other ATP-dependent chaperones (Liu 2015; Zhang 2015).

**Hsp60 Family**

The eukaryotic Hsp60 chaperone, mainly found in the mitochondria, is constitutively expressed under normal conditions and is induced by several types of stress, such as heat shock, oxidative stress, and DNA damage. Together with its cofactor Hsp10, Hsp60 is the main machinery for the correct folding of mitochondrial proteins. Moreover, Hsp60 has been found in other cellular compartments, such as the cytosol and plasma membrane; also, it may be exported to the extracellular space, through vesicle-mediated secretory pathway. In the extracellular environment, Hsp60 interacts with receptors present on immune cells and reach the bloodstream, where it acts as a chaperokine (a chaperone with cytokine activity) at distant places (Asea 2000). Furthermore, Hsp60 plays several important roles in some human diseases, being a reliable biomarker for following patients with temporal lobe epilepsy as well as for evaluating cardiovascular risk (Cheng 2014; Gammazza 2015).

In the bacteria *Escherichia coli*, the chaperonin GroEL (a 57 kDa protein) exists as a 14-meric cylinder consisting of two heptameric rings. Together with its co-chaperonin GroES (a 10 kDa protein), existing as a single 7-meric ring, GroEL favors protein folding through an ATP-dependent process. The molecular complex GroEL/GroES binds to a large variety of polypeptide substrates, ranging from 2 to 100 kDa. The apical domains of the GroEL subunits, forming the ring opening, expose the hydrophobic residues for binding of substrates. Upon ATP binding to GroEL, GroES covers the GroEL ring that holds the substrate, generating a closed compartment large enough to contain polypeptides of up to 60 kDa. In this conformation, the substrate is folded during the hydrolysis of seven molecules of ATP in the GroEL ring. The binding of ATP to the opposite ring causes the dissociation of GroES and the subsequent release of the substrate. If the substrate polypeptide has not yet been fully folded, it is recaptured for another round of folding inside GroEL (Georgescauld 2014).

The CCT chaperone, also called TRiC (TCP-1 ring complex), has been found in mammalian and yeast cells. To assist protein folding, sixteen CCT monomers assemble a protein complex, consisting of two stacked rings of eight subunits each. This complex forms two inner chambers where substrate polypeptides are trapped and then folded. CCT subunits are highly conserved; however, they diverge in the binding regions, allowing functional adaptation to a wide variety of substrates, including actin, tubulin, STAT3, and some cell cycle regulators. CCT is essential for de novo folding of about 10% of the newly synthesized polypeptides and, also, for refolding of proteins denatured by cellular stress. Interestingly, substrates for CCT are characterized by the presence of domains with a marked tendency to form β-sheets (Rüßmann 2012; Kasembeli 2014). CTT cycles between open and closed conformations, which are regulated by ATP. The closed conformation is favored by the transition state of ATP hydrolysis. The cavity shaped in CTT provides plenty space for 70 kDa substrate polypeptides; however, it has the ability to fold larger substrates. For the latter, CCT only captures the protein domains that requires folding assistance (Rüßmann 2012).

## Hsp70 Family

In bacteria, the DnaK chaperone (a prokaryotic homolog of Hsp70) plays a major role in misfolding of proteins, preventing aggregation, assisting refolding, and remodeling large aggregates (Evans 2012; Cho 2015). DnaK has two functional domains: the N-terminal ATPase domain (NBD) and the C-terminal substrate binding domain (SBD). The association or dissociation of the substrate is restricted by the nature of the nucleotide bound to the NBD. When ADP is bound (closed conformation), stable interactions with the substrate are favored, while the affinity is reduced when ATP is bound (open conformation) (Evans 2012; Doyle 2015). DnaK performs a wide variety of activities through association with co-chaperones or other chaperones. For instance, the ATPase cycle is controlled by the co-chaperones, DnaJ and GrpE. DnaJ (a member of the Hsp40 family) stimulates ATPase activity, whereas GrpE, a nucleotide exchange factor (NEF), promotes the release of ADP. Moreover, by association with NBD of DnaK, the ClpB chaperone (a member of the Hsp100 family) is activated, facilitating its ability to solubilize and reactivate aggregated proteins (Doyle 2015).

Other members of the Hsp70 family are the cytosolic Ssa and Ssb chaperones. Activated in response to a wide variety of stressor, the main function of Ssa proteins is to prevent aggregation of denatured polypeptides, by binding to their hydrophobic regions and preventing the generation of amorphous aggregates (Hasin 2014). Ssa proteins are also involved in protein translation, translocation, and folding. Moreover, cells Ssa-deficient exhibit multiple defects, including cell cycle arrest in $G_2$/M phase, accumulation of precursor proteins targeted to the endoplasmic reticulum (ER) or mitochondria, abnormal nuclear distribution, and aberrant microtubule formation. As other members of the Hsp70 family, Ssa proteins are associated with other co-chaperones, such as Ydj1 (a DnaJ homolog) and Fes1 (a NEF-like protein) (Shaner 2005).

In contrast, Ssb proteins are associated with both the ribosome and the nascent polypeptides emerging from the ribosomal exit channel. Cells lacking Ssb proteins exhibit enhanced sensitivity to inhibitors of translation, indicative of its role in the process. In addition, the ribosome-associated complex (RAC), a heterodimer comprised of Ssz-1 (Hsp70-like) and Zuo-1 (a J-protein), specifically activates the ATPase function of Ssb proteins (Shaner 2005). Therefore, it is likely that RAC and Ssb play key roles in cellular protection of toxic protein aggregates (Kiktev 2015).

A well-studied member of the Hsp70 family is the immunoglobulin heavy-chain-binding protein (BiP), an ER-resident chaperone that distinguishes and binds to unfolded regions of nascent polypeptides during translation. BiP is the main system for monitoring protein folding and participates in various critical processes, such as protein translocation, disaggregation, degradation (Behnke 2015). Also, BiP performs multiple cycles of binding and release of polypeptide substrates to promote proper folding.

This process is tightly regulated by a cycle of ATP binding and hydrolysis as well as by nucleotide exchange (Griesemer 2014; Behnke 2015). Furthermore, some cofactors regulating BiP activity, named ERdj proteins (similar to the bacterial DnaJ, located at the ER), directly bound to unfolded proteins and target this substrate to BiP. In addition, GRP170 and SiL1 (nucleotide exchange factors) facilitate the substrate release from BiP by stimulating the release of ADP and allowing ATP to rebind to the nucleotide binding domain, promoting a conformational change that favors the open state of BiP and allowing accessibility to the substrate binding domain (Araki 2011; Behnke 2015).

### Hsp90 Family

A well-studied Hsp90 family member is the Grp94 chaperone, a 94-kilodalton glucose regulated protein, which is the most abundant glycoprotein in the ER (Zhu 2015). Besides participating in protein folding, Grp94 interacts and associates with other chaperones as well as is involved in storage of calcium and assists in targeting of misfolded proteins to the mechanism of ER-associated degradation (ERAD) (Eletto 2010; Marzec 2012). Interestingly, Grp94 substrates are selected, some of which play critical roles in immunity, growth, signaling, and cell adhesion (e.g., major histocompatibility class I, insulin-like growth factor II, Toll-like receptor I, and a subset of integrins) (Liu 2008; Eletto 2010; Staron 2010). Also, Grp94 exhibits a surface and secreted forms that facilitate antigen presentation and immune response (Luo 2013).

Another Hsp90 family member is the mitochondrial TRAP1 chaperone (tumor necrosis factor receptor-associated protein 1), which protects the cell from apoptosis induced by reactive oxygen species (ROS) and senescence (Im, 2014). Recently, it has been showed that TRAP1 is involved in the association of hexokinase II to the mitochondrial outer membrane and promotes the folding and stability of the succinate dehydrogenase, a subunit of complex II from the oxidative phosphorylation chain (Lisanti 2014).

## 2.2. Calnexin and Calreticulin

Calnexin (CNX) and calreticulin (CRT) are ER lectin chaperones involved in folding and quality control of glycoproteins. CRT is a type I membrane protein, while CNX is its soluble paralog having 39% of sequence homology. CNX and CRT share similar structural features: both have the N-terminal globular domain containing the lectin binding site. In CNX, the globular domain is formed by a β-sandwich of concave and convex β-sheets (Tannous 2015).

In the ER compartment, glycoprotein folding, transport, and degradation is coupled to the N-glycosylation process. An oligosaccharide is transferred to asparagine residues in acceptor polypeptides. Then, glycosidase enzymes modify the branching pattern, leaving a mono-glycosylated glycan, which mediates binding to the ER lectins (CNX or CRT). The latter interaction allows ER-retention of misfolded substrates and, also, promotes refolding through interactions with foldases, such as peptidyl-prolyl cis-trans isomerase

(PPI) and protein disulfide isomerase (PDI) (Ferreira 2004; Banerjee 2007; Ferris 2013; Tannous 2015). Once the substrate is correctly folded, the glycoprotein is transported to the Golgi apparatus. Otherwise, it can either be retained to get another round of refolding or be directed to ERAD (Ferris 2013; Tannous 2015).

## 2.3. Peptidyl-Prolyl cis-trans Isomerases

Peptidyl-prolyl cis-trans isomerases (PPI) contribute to the folding and restructuring of substrate polypeptides by catalysis of the slow rotational motion of peptide bonds preceding a proline residue. PPI family include parvulins, cyclophilins, and FK-506 binding proteins (FKBP) (Schiene-Fischer 2014). All proteins have a three-dimensional structure where the configuration of each peptide bond is clearly defined. In correctly folded proteins, about 7% of proline bonds have a cis configuration, while non-proline bonds exhibit just 0.03% (Schmidpeter 2015). Hence, PPI promotes folding of partially folded polypeptides by catalyzing the cis-trans isomerization of proline residues with incorrect configuration (Schiene-Fischer 2014; Schmidpeter 2015).

## 2.4. Protein Disulfide Isomerases

Protein disulfide isomerases (PDI) are oxidoreductases with chaperone activity belonging to the superfamily of thioredoxin-like proteins. Although its primary function is to catalyze the formation and rearrangement of disulfide bonds of polypeptide substrates, PDI enzymes have shown the ability to function as chaperones, a process called oxidative folding of proteins (Stolf 2011; Biran 2014).

Even though they are ER-resident proteins, these enzymes have been found in the cytoplasmic membrane, where catalyze the reduction of disulfide bonds from surface proteins, enhancing adhesion and cell migration. Furthermore, this membrane-associated activity has also been implicated in several processes that regulate host-pathogen interaction during viral and parasitic infections (Santos 2009; Pan 2014).

## 3. Medically-Important Protozoan Parasites

Interestingly, a set of chaperones and foldases has been identified in numerous organisms studied so far, including eukaryotic parasites that cause human diseases. Worldwide, parasitic infections are prevalent and remain as a significantly threat to the public health. Tables 1 and 2 summarize the clinical pathology and infective stages of some medically-important protozoan parasites.

**Table 1. Medically-important intracellular protozoan parasites**

| Pathogen | Clinical pathology | Infective stages |
|---|---|---|
| Plasmodium | In endemic areas, patients with severe disease usually exhibit one or more of the following: severe anemia, respiratory distress associated with metabolic acidosis, or cerebral malaria. In some cases, patients can acquire partial immunity, allowing the emergence of asymptomatic infections (WHO – Malaria 2015). | The infective stage is named sporozoite. Within the host cell, sporozoites mature into schizonts, then, by asexual replication (schizogony), they release merozoites. When these infect erythrocytes, the ring stage trophozoite matures into schizonts. Some of the intra-erythrocytic trophozoites mature into sexual forms, called gametocytes. Blood stages are responsible for the clinical manifestations of the disease (CDC – Malaria 2012). |
| Leishmania | Two clinical manifestations can be associated to this parasite infection. Visceral leishmaniasis (VL), characterized by irregular fever, weight loss, hepatosplenomegaly, and anemia. Cutaneous leishmaniasis (CL), characterized by the occurrence of one or more lesions on the skin (face, trunk, or extremities), that can change size and appearance over time (WHO – Leishmaniasis 2015). | The extracellular infective stage, promastigote, develops and multiplies in the digestive tract of the transmitting insects (vectors). The intracellular replicative stage, amastigote, resides and multiplies within the phagolysosomes of the host mononuclear phagocytes (CDC – Leishmaniasis 2013). |
| Trypanosoma | The lesions produced by the infection of *T. cruzi* (Chagas disease) depend on both parasite and host. Three clinical phases can be distinguished. Acute phase, asymptomatic in about 70% of those infected, characterized by high parasitaemia and multiple tissue invasion. Indeterminate phase, without detectable parasitaemia, where anatomical and functional abnormalities have been reported, and also sudden death. Chronic phase, characterized by irreversible visceral damage (WHO – Chagas 2015). | Trypomastigotes enter the host through a wound or mucosal membranes. Inside, these can invade nearby cells, where they differentiate into amastigotes. These intracellular forms then multiply and differentiate into trypomastigotes, which are released to the bloodstream and spread the infection to other cells and tissues. Within the vector, these trypomastigotes are transformed into epimastigotes, a mobile and replicative stage (CDC – American Trypanosomiasis 2015). |

**Table 2. Medically-important protozoan parasites associated with diarrheic diseases**

| Pathogen | Clinical pathology | Infective stages |
|---|---|---|
| Giardia | A multifactorial process is associated to the infection with Giardia, which involve functional and immunological aspects of host and parasite. The clinical manifestation range from asymptomatic to acute or chronic disease. The diarrheic syndrome appears as a common acute manifestation (Halliez 2013). | Two stages are clearly distinguished: the cyst, infective form, resistant to the environment outside the host, and the trophozoite, the invasive form, which colonize the small intestine of the host (Halliez 2013; CDC – Giardia 2010). |
| Entamoeba | The intestinal infection (acute and chronic) is characterized by abdominal pain, tenesmus, and sometimes acute dysenteric diarrhea. The toxic colitis is generated by the invasion and perforation of the colon, which creates a non-specific toxic situation, causing severe peritonitis. The disseminated infection is characterized by invasion of other organs, forming abscesses, the most frequently affected is the liver (Ralston 2011). | The cyst stage, infective form, is resistant to external environmental conditions and can survive in the gastric juice. The trophozoite stage, invasive form, has the ability to attach to the mucosa and colonize the intestinal wall (Fletcher 2012; CDC – Amebiasis 2010). |

Protozoan parasites of the phylum Apicomplexa are the most prevalent in the world (Frölich 2011). Within this, parasites of the genus Plasmodium are the most common are the, which cause malaria and are transmitted to humans through the bites of infected Anopheles mosquitoes. There are four species that cause malaria: *P. falciparum, P. vivax, P. malariae* and *P. ovale.* According to the latest global estimates, in 2013, there were roughly 198 million cases of malaria that led to the death of about 584,000 people (WHO – Malaria 2015).

Leishmaniasis is caused by a protozoan parasite of the genus Leishmania, which includes more than 20 different species, and it is transmitted to humans through the bite of infected female sandflies. The disease occurs in three main forms: visceral (the most severe), cutaneous (the most common), and mucocutaneous. Annually, 1.3 million new cases and between 20,000 and 30,000 deaths are estimated (WHO – Leishmaniasis 2015; Alvar 2012).

Chagas disease, also known as American Trypanosomiasis, is caused by the protozoan parasite *T. cruzi.* It is mostly vector-transmitted to humans by contact with feaces of Triatominae bugs. It has been estimated that between 6 and 8 million people are infected worldwide, most of them in Latin America, where Chagas disease is endemic. Moreover, a total of 56,000 new cases have been reported annually, causing about 12,000 deaths (WHO – Chagas 2015).

WHO reported in 2014 that diarrheal diseases affected more individuals than any other disease associated with infections (WHO – Statistics 2014). In humans, these diseases are caused by different pathogens, including protozoan parasites. *Giardia lamblia*, causative agent of giardiasis, is the most common enteric protozoa with a cosmopolitan distribution (Fletcher 2012). *Entamoeba histolytica*, causative agent of amebiasis, is also a cosmopolitan protozoa, but more prevalent in developing countries with poor sanitary conditions. According to WHO, amebiasis is the third leading cause of death by parasitic infections (WHO/PAO/UNESCO 1997).

## 4. Proteostasis and Proteotoxic Stress in Protozoa

Numerous and complex biological processes simultaneously occur in living cells. These processes must be strictly regulated to ensure an optimum balance in their functions. Proteins perform a wide variety of cellular functions, including the replication of genetic material, catalysis and metabolism, cell signaling, cell cycle, and membrane transport, among others. Hence, the life of a cell depends on the efficient function of thousands of proteins, which, in turn, depend on correct folding (Cuanalo-Contreras 2013).

Since proteostasis controls protein concentration, subcellular localization, chaperone-assisted folding, and protein degradation, this mechanism is responsible for maintaining the dynamic balance of the cell proteome (Gidalevitz 2010). Therefore, unbalanced proteostasis can lead to aberrant folding, aggregation and accumulation of proteins, causing stressful conditions that may result in cellular dysfunction (Bar-Lavan 2013). Some cells have a high rate of protein synthesis, which makes them more likely to accumulate misfolded proteins and to potentially develop proteotoxic stress (Kawabata 2012).

In eukaryotic cells, the accumulation of unfolded proteins in the ER induces proteotoxic stress, which then triggers a quality control mechanism, the unfolded protein response (UPR) (Hartley 2010). When UPR is efficient, cells can temporarily adapt to that stress (Papa 2012). However, excessive and prolonged ER stress induces cell death, usually by apoptosis, which is the last resort to dispense dysfunctional cells (Xu 2005). Chaperones and foldases that enhance folding capacity, and initiating the process to export and degrade misfolded proteins. The most abundant and best characterized ER chaperones and foldases include BiP, GRP94, CRT, and PDI. Interestingly, the increased protein synthesis in highly proliferative or secretory cells requires increased ER capacity and function. Therefore, ER chaperones and foldases play important roles in maintaining proteostasis and contributing to cell growth and survival (Hartl 2009; Luo 2013).

Because of ER connectivity to other subcellular compartments, it is possible to comprehend how some chaperones and foldases may be located outside such compartment. For instance, on the cell surface, where they participate in signaling functions that regulate proliferation and cell death (Peters 2011). Furthermore, different studies have shown the potential to regulate ER proteostasis, including the development of specific therapeutics aimed at producing proteotoxic stress in the ER (Liu 2011).

As many other organisms, protozoa also depend on mechanisms that promote efficient protein folding (Becker 1996). Furthermore, for pathogens such as Plasmodium, Trypanosoma, Leishmania, Toxoplasma, Entamoeba, and Giardia, it is essential that virulence factors exported to the plasma membrane or extracellular environment exhibit proper native structure, which is important to fulfill their functions (Joyce 2013; Sibley 2013). Although more information on protein folding in protozoa is needed, the general knowledge on protein homeostasis can be used to design strategies aimed to identify therapeutic molecules with anti-pathogenic activity; specifically, targeting key components of the protein folding machinery, leading to sustained proteotoxic stress, and thus inducing cell death (Calamini 2011; Calamini 2012).

## 5. Role of Protein Folding in the Life Cycle of Protozoan Parasites

During the course of its life cycle, protozoan parasites advance through different stages that involve morphological and physiological changes, allowing them to successfully perform several processes, such as infecting and invading the host, and surviving harsh environmental conditions (Florence 2002). This developmental adaptation clearly suggests an evolutionary selection that involves the acquisition of nutrients, release of virulence factors, microbial resistance, and effective immune evasion (Khalaf 2012).

Interestingly, the process of infection and invasion exposes the parasite to adverse environmental conditions, such as abrupt changes in temperature and pH, which may affect the stability of proteins (Acharya 2007). Therefore, the assistance of chaperones and foldases is essential to preserve protein homeostasis during the parasitic lifestyle of protozoa.

## 5.1. Plasmodium

Plasmodium parasites have a complex life cycle that includes several stages (Table 1). Complete cycle is described in greater detail elsewhere (Enomoto 2012; CDC – Malaria 2012; WHO – Malaria 2015). During transmission from the vector to humans, as well as in development within the human cell, the parasite has to confront unfavorable environmental conditions, such as heat stress, e.g., because the temperature in the vector and the human is different, and due to fever episodes in infected humans (Acharya 2007). Chaperones from The Hsp family allow parasite surviving under those conditions, by binding to denatured polypeptides and assisting proper refolding (Shonhai 2007).

The *P. falciparum* Hsp90 homolog (*Pf*Hsp90) plays two major roles in the parasitic lifestyle: i) folding of essential proteins, e.g., virulence factors, and preventing the formation protein aggregates (along with other members of the Hsp family) (Bonnefoy 1994), and ii) modulating the activity of some transcription factors (Banumathy 2003; Kumar 2003). Use of Hsp90 inhibitors has been key to elucidate its participation during infection and development of *P. falciparum*. When *Pf*Hsp90 is inhibited, a large number of Hsp90-substrate polypeptides, which are necessary for development of the parasite within the human erythrocyte, fail to fold properly and undergo degradation (Banumathy 2003; Kumar 2003). Interestingly, this inhibition arrested the parasite progress at an immature stage. However, this condition had no effect on the release of mature forms, which were able to efficiently infect fresh erythrocytes (Banumathy 2003).

Another Hsp-like protein, *Pf*Hsp70, participates in the parasite development during heat stress. The expression of *Pf*Hsp70 is particularly notable in the intra-erythrocyte stages, where is involved in the folding of secreted proteins, granting the settlement of the parasite within the cell (Shonhai 2007). In addition, *Pf*Hsp70 and *Pf*Hsp90, along with some co-chaperones, constitute a multi-chaperone complex engaged in the folding of several proteins (Acharya 2007; Gitau 2012). Moreover, another complex, formed by *Pf*Hsp70 and *Pf*Hsp40 (a co-chaperone), can function as a refoldase during protein trafficking between the parasite and the erythrocyte (Birkholtz 2008). It seems that *Pf*Hsp40 is important for maturation of the parasite, as the co-chaperone is particularly abundant in this life cycle phase (Achayra 2007).

A large number of proteins involved in virulence and pathogenesis of Plasmodium parasites have cysteine residues forming disulfide bonds, e.g., the apical membrane antigen 1 (AMA-1) or the EGF-like domain of the merozoite surface protein 1 (MSP-1) (Haque 2012). Both proteins AMA-1 and MSP1 depend on accurate oxidative folding to efficiently perform its function (Hodder 1996; Guevara 1997; Moss 2012). In eukaryotic cells, disulfide bond formation is catalyzed mainly by PDI enzymes. In the genus Plasmodium, nine PDI homologs have been described so far (four paralogs in *P. falciparum*) (Haque 2012). PDI enzymes are highly expressed in asexual stages of this parasite (Mahajan 2006). Moreover, it has been shown that oxidative folding of the erythrocyte binding antigen 175 (EBA-175), a virulence factor and leading malaria vaccine candidate, is assisted by a *Pf*PDI enzyme (Gilberger 2003; Mahajan 2006).

## 5.2. Leishmania

The life cycle of Leishmania parasites comprises two major stages (Table 1). Complete cycle is described in greater detail elsewhere (CDC – Leishmaniasis 2013; WHO – Leishmaniasis 2015). The parasite has developed unique adaptive mechanisms to ensure survival in the harsh environments faced throughout their life cycle (Cunningham 2002). Moreover, to avoid denaturation of proteins, the parasite contains a set of chaperones and foldases playing an important role to ensure successful infection and invasion processes. Within these, some members of the Hsp family have been identified as key in the cellular response to heat shock, by preventing denaturation and aggregation of proteins during infection (Teixeira 2015).

In *L. brazilensis*, the Hsp70 homolog has been reported as one of the major chaperone involved in prevention of aggregation and folding assistance of proteins (Ramirez 2013). Moreover, its biological significance has been confirmed by gene disruption, in *L. intantum* (Folgeira 2008), and knockdown, in *L. donovani* (Raina 2012), leading to defects on cell cycle, growth, morphology, and virulence of the parasite. Interestingly, it seems that thermal stress is required as a signal for parasite differentiation (Clos 1999). Furthermore, it has been suggested that the transient increase in the expression of heat shock proteins, which precedes the transformation between stages, is associated with the extensive remodeling that accompanies the morphologic transition (Miller 2000).

The Hsp100 and Hsp90 homologs are two additional chaperones involved in Leishmania differentiation. When the parasite is stressed, high levels of Hsp100 are detected (Krobitsch 1999), functioning as integrity protector. Although not decisive for differentiation, it has been shown that Hsp100 is relevant in maintenance of the intracellular stage, as it acts keeping correctly folded essential proteins and some virulence factors (Krobitsch 1999; Clos 2001). On the contrary, Hsp90 is essential for stage differentiation (Wiesgigl 2001). Recently, two isoforms of the p23 co-chaperone has been identified in *L. brazilensis* (Batista 2015). This co-chaperone binds to the Hsp90-ATP complex. Despite having similar affinities for Hsp90, the two isoforms have dissimilar thermal stability profiles and chaperone activities, suggesting that each might be required in different stages of the parasite life cycle.

On the other hand, *L. major* contains a PDI homolog that plays a key role on its parasitic lifestyle. Mutants lacking *Lm*PDI exhibit a non-virulent phenotype, indicating that it may participate in attachment, internalization, or multiplication (Markikou-Ouni 2012). Moreover, strong suppression of parasite proliferation was observed in cultures treated with bacitracin, a known PDI inhibitor (Khalaf 2012). Furthermore, it has been shown that PDI function is crucial for efficient infection of *L. chagasi* to macrophages (Santos 2009). Also, in *L. donovani*, the CRT homolog is important for the targeting of some virulence factors during their traffic through the parasite secretory pathway. Interestingly, altering the chaperone function of CRT affects secretion of acid phosphatases, resulting in a significant decrease in the survival of L. donovani within human macrophages (Debrabant 2002).

## 5.3. Trypanosoma

Trypanosoma parasites also have a complex life cycle which includes several stages (Table 1). Complete cycle is described in greater detail elsewhere (Nagajyothi 2012; Michaeli 2015; CDC – American Trypanosomiasis 2015; WHO – Chagas 2015). As expected, during transmission and infection, the parasite must be able to withstand harsh environmental conditions and survive the attacks of the immune system, among others. To understand the role of chaperones in the life cycle of Trypanosoma, thermal or ER stress has been used to study the role of several protein homologs, including Hsp, CRT, and cyclophilin (Paba 2004; Michaelli 2015).

In *T. cruzi*, the Hsp90 chaperone is not involved in stage differentiation. However, inhibition of *Tc*Hsp90 induces a stress response and causes growth arrest, indicating that protein folding is needed to complete the cell cycle of the parasite (Graefe 2002). Additionally, this parasite has four Hsp70 isoforms that are differentially expressed during its life cycle (Shonhai 2007). Furthermore, it has found that some typical chaperone complexes are formed, such as Hsp70-Hsp40 (Edkins 2004).

A major role of protein oxidative folding in Trypanosoma parasites is illustrated by the variable surface glycoproteins (VSG), which are essential for evading the host immune system and maintaining the low level of chronic infection, since they contain several disulfide bonds that are important for structural integrity (Haque 2012). Specific chaperones and foldases responsible for proper folding of VSG remains to be revealed; however, the implication of several Hsp homologs, some PDI enzymes, and the lectin CRT has been proposed (Wang 2010; Field 2010).

## 5.4. Giardia and Entamoeba

The life cycle of the enteric parasites Giardia and Entamoeba is quite simple (Table 2). Complete cycles are described in greater detail elsewhere (Ralston 2011; Fletcher 2012; Halliez 2013; CDC – Amebiasis 2010; CDC – Giardia 2010). Interestingly, both parasites require the assistance of chaperones and foldases during the synthesis of essential proteins and virulence factors (Ratner 2008; Ramos 2011; Haque 2012).

Cellular functions of the Hsp90 chaperone are associated with parasite development (Debnath 2014). The *G. lamblia* homolog (*Gl*Hsp90) is encoded by two fragments, separated by a large stretch of sequence, which are spliced together in trans (Nageshan 2011; Roy 2012). Apparently, this features helps the parasite to elude natural inhibitors (Roy 2012). Little is known regarding the cellular role of Hsp90 in *E. histolytica* (*Eh*Hsp90): exhibits a weak ATPase activity and is capable of binding to *Eh*Aha-1 (a co-chaperone). Also, it seems that *E. histolytica* is dependent on *Eh*Hsp90 for growth and survival (Singh 2014).

On the other hand, both Giardia and Entamoeba parasites have PDI enzymes involved in protein folding (Haque 2012). In Giardia, two major proteins of the cell wall are targeted to the secretory pathway and are covalently associated by disulfide bonding (Hehl 2000). Furthermore, PDI activity is crucial in the ER during the differentiation of Giardia parasites. For instance, when disulfide bond formation is inhibited with dithiothreitol, a membrane-permeable reductant, parasite differentiation is irreversible blocked (Knodler 1999). Also, PDI enzymes play an essential role in protein folding of Entamoeba parasites. In *E. histolytica*, some virulence factors that are essential for its parasitic lifestyle require proper oxidative folding to perform their cellular function (Hecht 2004; Chatterjee 2010). Moreover, using *E. invadens* as a model to study the life cycle of Entamoeba parasites, it has been shown that two cell wall proteins, both containing chitin-binding domains stabilized by disulfide bonding, are essential for stage transition (Samuelson 2011). Hence, it has been proposed that a PDI enzyme might catalyze the disulfide boding in amebic proteins involved on its life cycle (Ramos 2008; Mares 2009; Ramos 2011; Mares 2015).

Other foldases, such as cyclophilin and PPI, have been identified in Giardia and Entamoeba parasites. Although participation of cyclophilin in the proliferation of *E. histolytica* has been reported, the specific cellular role in both parasites remains to be elucidated (Ostoa-Saloma 2000; Buchko 2013).

## 6. Molecular Chaperones As Anti-protozoan Drug Targets

As described so far, protozoan parasites should adapt to different environments throughout their life cycle. They must survive several life-threatening conditions, including sudden changes in pH and temperature as well as different oxygen tension and low-nutrient situations. Also, they have to deal with oxidative stress and immune response in their host. Hence, to resist and survive stressful conditions and complete their life cycle, protozoa have developed a robust molecular machinery, which includes chaperones and foldases as key components (Neckers 2008). Moreover, the potential of these components as targets for the discovery or development of new anti-parasitic is evident (Wiesgigl 2001; Graefe 2002; Kumar 2003; Kumar 2007; Jones 2008; Li, 2009; Mout 2012; Angel 2013; Meyer 2013; Santos 2014).

Among the parasite chaperones and foldases, Hsp90 chaperone has an exceptional potential as a drug target. In protozoan parasites, several inhibitors targeting functions of the Hsp90 counterpart have been reported (Table 3), being geldanamycin (GA), an antibiotic produced by *Streptomyces hygroscopicus*, the most studied.

**Table 3. Inhibitors of Hsp90 homologs from medically-important protozoan parasites**

| Organism | Inhibitor | Mode of action |
|---|---|---|
| *P. falciparum* | Geldanamycin | Inhibits the ATPase activity. Blocks some stage transitions during development (Kumar 2003; Angel 2013). |
| | 17-AAG(1)<br>17-PEG-Alkyn-GA(2) | Exhibit affinity for the ATP binding domain, inhibiting its ATPase activity (Kumar 2007; Mout 2012). |
| | Harmine | Exhibits selectivity for the ATP binding site. Inhibits early stage transitions (Shahinas 2012). |
| | PU-H71 | Exhibits high affinity for the ATP binding site, inhibiting its ATPase activity (Shahinas 2013). |
| *L. donovani* | Geldanamycin | Inhibits the ATPase activity. Induces cell cycle arrest. Promotes an apoptotic death (Wiesgigl 2001; Li, 2009). |
| | Radicol | Exhibits affinity for the ATP binding site. Arrest the cell growth (Wiesgigl 2001). |
| *L. brazilensis* | Geldanamycin | Inhibits the ATPase activity. Promotes degradation rather than folding (Silva 2013). |
| | 17-AAG(1) | Inhibits the ATPase activity, Promotes arrest of the cell growth (Santos 2014). |
| *L. amazonensis* | 17-AAG(1) | Inhibits the ATPase activity. Affects the viability of the parasite in the early phases of treatment (Petersen 2012). |
| *T. brucei* | Geldanamycin | Exhibits affinity for the ATP binding site. Reduces the cell growth in specific stages (Jones 2008; Meyer 2013). |
| | Radicicol | Exhibits affinity for the ATP binding site. Inhibits the ATPase activity (Meyer 2013). |
| | Novobiocin | Exhibits affinity for the C-terminal domain. Induces conformational changes in the chaperone structure (Allan 2006; Meyer 2013). |
| | 17-AAG(1)<br>17-PEG-Alkyn-GA(2)<br>17-AG(3)<br>17-DMAG(4) | Exhibit affinity for the ATP binding domain. Inhibits the ATPase activity. Promotes arrest of growth of specific stages (Jones 2008; Meyer 2013). |
| *T. cruzi* | Geldanamycin | Exhibits affinity for the ATP binding domain. Inhibits the ATPase activity. Inhibits cell growth and prevents cell differentiation (Graefe 2002). |
| *G. lamblia and E. histolytica* | SNX-2112 and similar | Compete with ATP for the nucleotide binding site (Debnath 2014). |
| *E. histolytica* | Geldanamycin | By molecular modeling, exhibits affinity for the chaperone (Singh 2014). |
| | 17-AAG(1) | Exhibits affinity for the ATP binding site. Inhibits the ATPase activity (Singh 2014). |

(1) 17-Allylamino-17-demethoxygeldanamycin.

(2) 17-N-(3-(2-(2-(3-aminopropoxy)ethoxy)ethoxy)propyl)-pent-4-ynamide-17-demethoxygeldanamycin.

(3) 17-(amino)-17-demethoxygeldanamycin.

(4) 17-(dimethylaminoethylamino)-17-demethoxygeldanamycin.

GA binds to the adenosine nucleotide-binding domain of Hsp90, competing with ATP for the binding site; thus, inhibiting the hydrolysis of ATP. Also, GA prevents Hsp90 substrate proteins from being correctly folded, and consequently promoting their degradation. By these actions, GA disrupts critical protozoan cell functions, such as growth and proliferation (Li 2009; Meyer 2013; Debnath 2014; Santos 2014). Although GA inhibits Hsp90 through the same mechanism, their cellular effects may vary in different protozoan parasites (Table 3).

In mammalian cells, the $IC_{50}$ of GA for arresting cell growth is roughly 350 μM (Jones 2008). Whilst in protozoan parasites, the drug affects growth in the nM range. This demonstrates the effectiveness of GA as a therapeutic agent for treatment of parasitic infections (Pallavi 2010; Meyer 2013, Singh 2014; Santos 2014). However, the clinical use of GA was not approved, due to the potential risk of hepatotoxicity. Therefore, efforts have been made to obtain GA derivatives with reduced toxicity, while retaining or improving its capacity as anti-parasitic drugs (Kumar 2007; Jones 2008; Mout 2012; Meyer 2013; Li 2015).

Another Hsp chaperone with potential as a drug target is Hsp70. Contrary to Hsp90, few studies have been reported regarding Hsp70 inhibitors targeting the protozoan counterparts. In *P. falciparum*, the marine alkaloid malonganenone A was identified as modulator of *Pf*Hsp70-1 and *Pf*Hsp70-x activities (Cockburn 2011; Cockburn 2014).

Particularly, the *Pf*Hsp70-x sequence contains unique substitutions, often in regions highly conserved in other orthologs, which make it a potential target for anti-malaria drug design (Hatherley 2014).

The ER-resident CRT lectin may also be regarded as a potential target for development of novel anti-parasitic agents. In *T. cruzi*, the CRT counterpart (*Tc*CRT) plays a key role in evasion of the host immune system (Ferreira 2004; González 2015). Similarly, the *E. histolytica* homolog (*Eh*CRT) helps the parasite in its ability to evade the immune system. Specifically, *Eh*CRT inhibits the classical complement pathway, thus protecting the pathogen from host-induced lysis (Ximénez 2014).

The PDI enzymes, a family of foldases, also have potential as targets for the design of new therapeutic agents (Stolf 2011). In *P. falciparum*, eight PDI enzymes have been identified. One of them, P*f*PDI-8, is highly similar to typical PDI enzymes, structurally and functionally. Moreover, it has been shown that some family members are expressed in all developmental stages while other are stage specific, suggesting that a small subset of enzymes, even just one of them, fulfills the main functions of this class of chaperones in the life cycle of the parasite (Mahajan 2006). In *T. brucei*, two PDI enzymes have been characterized. Although these are not essential for the growth, they play an important role in the defense against the host immune system, by reducing disulfide bonds or by unfolding internalized antibodies to promote rapid proteolysis (Rubotham 2005). In Leishmania parasites, five PDI enzymes were identified in both *L. major* and L. amazonensis (Stolf 2011). Moreover, it has been reported that typical PDI inhibitors, such as bacitracin and *p*CMBA (*p*-chloromercuribenzoic acid), block the enzymatic activity and the parasite growth in culture, suggesting that PDI foldases play a role in the Leishmania development (Hong 2008). Moreover, a PDI foldase from Leishmania has been considered as an antigen for development of vaccines against parasitic infection (Benhnini 2009). In *E. histolytica*, a PDI family of eleven enzymes has been reported (Ramos 2008). One of them, *Eh*PDI, is actively expressed in the infective stage of the parasite, and exhibits the structural features of a typical P5 polypeptide and the functional activities of a classic PDI oxidoreductase (Ramos 2005; Mares 2009; Ramos 2011; Mares 2015). Furthermore, it has been shown that bacitracin inhibits the oxidoreductase activities of *Eh*PDI, suggesting that blocking the function of this foldase can be used as a target for development of alternative anti-amebic drugs (Ramos 2011).

PPI activity has been recognized in three classes of enzymes: cyclophilins, FK-506 binding proteins (FKBP), and parvulins. Cyclophilins are inhibited by cyclosporins and sanglifehrins, FKBP by FK-506 and its analogs, and parvulins by juglone (Lawen 2014). Protozoan parasite counterparts of cyclophilins and FKBP are essential for their life cycle. The anti-parasitic effect of cyclosporine A (CsA) was reported in *E. histolytica*. A dose-dependent inhibition of cell growth was observed when CsA was supplemented to the culture medium. Although the anti-amebic mechanism is not certain, the drug might be inhibiting the isomerase activity of the amebic PPI homolog (*Eh*Cyp) (Ostoa-Saloma 2000). CsA has also a significant inhibitory effect of CsA on the growth of *L. donovani*. This effect clearly suggests an important role for PPI enzymes in the parasite biology. The suppressive action on the host immunity disregards its use as an anti-parasitic drug. However, future research must be intended to identify novel inhibitors that specifically target the PPI enzyme of the parasite (Yau 2010; Yau 2014). CsA also exhibits anti-parasitic activity in *P. falciparum*. Cyclophilins play an important role in the development of this parasite, promoting the export of virulence factors and acting as chaperone in response to heat shock conditions (Marín-Menéndez 2011). Although the anti-parasitic effect of PPI inhibitors has not been studied in *G. lamblia*, a FKBP-like enzyme (*Gl*FKBP) was already identified (Buchko 2013), supporting the notion that PPI enzymes can be targets for designing drugs with anti-giardial activity.

## Conclusion

Proteostasis is essential for the proper functioning of any cell, including protozoan parasites. These pathogens have a set of molecular chaperones and foldases that are involved in both the maintenance of the intracellular protein balance and the resistance to adverse environmental conditions (inside and outside their hosts). Also, some of them have been associated with pathogenesis, as they favor the correct folding of important virulence factors, a process that is critical for infection. Although the first studies were conducted more than ten years ago, the search for specific inhibitors directed to block the function of molecular chaperones or foldases from protozoan parasites remains as productive area. Furthermore, the ultimate goal of drug discovery is the rational design of parasite-specific chemotherapeutics. Finally, it is important to highlight that targeting any molecular chaperone or foldase could lead to a success in the drug therapy, since those molecules are located in the center of several protein networks, with diverse functions, which allow the development, establishment, and surviving of the pathogen in adverse environmental conditions.

## References

Acharya P, Kumar R, Tatu U. Chaperoning a cellular upheaval in malaria: Heat shock proteins in *Plasmodium falciparum*. *Mol. Biochem. Parasitol.* 2007; 153 (2): 85-94.

Allan RK, Mok D, Ward BK, Ratajczak T. Modulation of chaperone function and cochaperone interaction by novobiocin in the C-terminal domain of Hsp90: evidence that coumarin antibiotics disrupt Hsp90 dimerization. *J Biol Chem.* 2006; 281 (11): 7161-71.

Alvar J, Vélez ID, Bern C, Herrero M, Desjeux P, Cano J, et al. Leishmaniasis worldwide and global estimates of its incidence. *PLoS ONE.* 2012; 7 (5): e35671.

Angel SO, Matrajt M, Echeverria PC. A review of recent patents on the protozoan parasite HSP90 as a drug target. *Recent Pat. Biotechnol.* 2013; 7 (1): 2-8.

Araki K, Nagata K. Protein Folding and Quality Control in the ER. *Cold Spring Harb. Perspect. Biol.* 2011; 3: a007526.

Asea A, Kabingu E, Stevenson MA, Calderwood SK. HSP70 peptide-bearing and peptide-negative preparations act as chaperokines. *Cell Stress Chaperones.* 2000; 5 (5): 425-31.

Banerjee S, Vishwanath P, Cui J, Kelleher DJ, Gilmore R, Robbins PW, Samuelson J. The evolution of N-glycan-dependent endoplasmic reticulum quality control factors for glycoprotein folding and degradation. *Proc. Natl. Acad. Sci. U.S.A.* 2007; 104 (28): 11676-81.

Banumathy G, Singh V, Pavithra S, Tatu U. Heat shock protein 90 function is essential for *Plasmodium falciparum* growth in human erythrocytes. *J. Biol. Chem.* 2003; 278 (20): 18336-45.

Bar-Lavan Y, Kosolapov L, Frumkin A, Ben-Zvi A. Regulation of cellular protein quality control networks in a multicellular organism. *FEBS J.* 2012; 279 (4): 526-31.

Batista F, Almeida G, Seraphim T, Silva K, et al. Identification of two p23 co-chaperone isoforms in *Leishmania brazilensis* exhibiting similar structures and Hsp90 interaction properties despite divergent stabilities. *FEBS J.* 2015; 282 (2): 388-406.

Becker B, Melkonian M. The secretory pathway of protists: spatial and functional organization and evolution. *Microbiol. Rev.* 1996; 60 (4): 697-721.

Behnke J, Feige MJ, Hendershot LM. Bip and its nucleotide exchange factors grp170 and sil1: mechanisms of action and biological functions. *J. Mol. Biol.* 2015; 427: 1589–1608.

Benhnini F, Chenik M, Laouini D, Louzir H, et al. Comparative evaluation of two vaccine candidates against experimental leishmaniasis due to *Leishmania major* infection in four inbred mouse strains. *Clin. Vaccine Immunol.* 2009; 16 (11): 1529-37.

Biran S, Gat Y, Fass D. The Eps1p protein disulfide isomerase conserves classic thioredoxin superfamily amino acid motifs but not their functional geometries. *PLoS ONE.* 2014; 9 (12): e113431.

Birkholtz L, Blatch G, Coetzer T, Hoppe H, et al. Heterologous expression of plasmodial proteins for structural studies and functional annotation. *Malaria J.* 2008; 7: 197.

Bonnefoy S, Attal G, Langley G, Tekaia F, Mercereau-Puijalon O. Molecular characterization of the heat shock protein 90 gene of the human malaria parasite *Plasmodium falciparum*. *Mol. Biohem. Parasitol.* 1994; 67: 157-170.

Brüning A, Jückstock J. Misfolded proteins: from little villains to little helpers in the fight against cancer. *Front Oncol*. 2015; 5: 47.

Buchko GW, Hewitt SN, Van Voorhis WC, Myler PJ. Solution structure of a putative FKBP-type peptidyl-propyl cis-trans isomerase from *Giardia lamblia*. *J. Biomol* NMR. 2013; 57 (4): 369-74.

Calamini B, Morimoto RI. Protein homeostasis as a therapeutic target for diseases of protein conformation. *Curr. Top Med. Chem.* 2012; 12 (22): 2623-40.

Calamini B, Silva MC, Madoux F, Hutt DM, et al. Small-molecule proteostasis regulators for protein conformational diseases. *Nat. Chem. Biol.* 2011; 8 (2): 185-96.

Caniparoli L, O´Brien EP. Modeling the effect of codon translation rates on co-translational protein folding mechanisms of arbitrary complexity. *J. Chem. Phys*. 2015; 142 (14): 145102.

CDC, Centers for Disease Control and Prevention. January 2013. Leishmaniasis. http://www.cdc.gov/parasites/leishmaniasis/biology.html (Accessed 06.26.2015).

CDC, Centers for Disease Control and Prevention. March 2015. American Trypanosomiasis. http://www.cdc.gov/parasites/chagas/biology.html (Accessed 06.26.2015).

CDC, Centers for Disease Control and Prevention. November 2010. Amebiasis. http://www.cdc.gov/parasites/amebiasis/biology.html (Accessed 06.26.2015).

CDC, Centers for Disease Control and Prevention. November 2010. Giardia. http://www.cdc.gov/parasites/giardia/biology.html (Accessed 06.26.2015).

CDC, Centers for Disease Control and Prevention. November 2012. Malaria. http://www.cdc.gov/malaria/about/biology/ (Accessed 06.26.2015).

Chatterjee A. Carpentieri A, Ratner DM, Bullitt E, et al. Giardia cyst wall protein 1 is a lectin that binds to curled fibrils of the GalNAc homopolymer. *PLoS Pathog*. 2010; 6 (8): e1001059.

Cheng YF, Sun JR, Chen HB, Abdelnasir A, et al. Association of Hsp60 expression with damage to rat myocardial cells exposed to heat stress *in vivo* and *in vitro*. *Genet. Mol. Res*. 2014; 13 (4): 9371-81.

Cho Y, Zhang X, Pobre KFR, Liu Y, et al. Individual and collective contributions of chaperoning and degradation to protein homeostasis in *E. coli*. *Cell Reports*. 2015; 11: 321–33.

Clos J, Klaholz L, Kroemer M, Krobitsch S, Lindquist S. Heat shock protein 100 and the amastigote stage-specific A2 proteins of *Leishmania donovani*. *Med. Microbiol. Immunol.* 2001; 190 (1-2): 47-50.

Clos J, Krobitsch S. heat shock as a regular feature of the life cycle of Leishmania parasites. *Amer. Zool.* 1999; 39 (9): 848-56.

Cockburn IL, Boshoff A, Pesce ER, Blatch GL. Selective modulation of plasmodial Hsp70s by small molecules with antimalarial activity. *Biol. Chem.* 2014; 395 (11): 1353-62.

Cockburn IL, Pesce ER, Pryzborski JM, Davies-Coleman MT, et al. Screening for small molecule modulators of Hsp70 chaperone activity using protein aggregation suppression assays: inhibition of the plasmodial chaperone *Pf*Hsp70-1. *Biol. Chem.* 2011; 392 (5): 431-8.

Cuanalo-Contreras K, Mukherjee A, Soto C. Role of protein misfolding and proteostasis deficiency in protein misfolding diseases and aging. *Int. J. Cell Biol.* 2013; 2013: 638083.

Cunningham A. Parasitic adaptive mechanisms in infection by Leishmania. *Exp. Mol. Pathol.* 2002; 141: 132-141.

Debnath A. Shaninas D, Bryant C, Hirata K, et al. *Hsp90 Inhibitors as new leads to target parasitic diarrheal diseases.* 2014; 58 (7): 4138-44.

Debrabant A, Lee N, Pogue GP, Dwyer DM, Nakhasi HL. Expression of calreticulin P-domain results in impairment of secretory pathway in *Leishmania donovani* and reduced parasite survival in macrophages. *Int. J. Parasitol.* 2002; 32 (11): 1423-34.

Doyle SM, Shastry S, Kravats AN, Shih Y-H, et al. Interplay between *E. coli* DnaK, ClpB and GrpE during protein disaggregation. *J. Mol. Biol.* 2015; 427: 312–27.

Edkins A, Ludewig M, Blatch G. A *Trypanosoma cruzi* heat shock protein 40 is able to stimulate the adenosine triphosphate hydrolysis activity of heat shock protein 70 and can substitute for a yeast heat shock protein 40. *Int. J. Biochem. Cell Biol.* 2004; 36 (8): 1585-98.

Efimov AV, Boshkova EA. Two Mechanisms of Protein Folding: A Theoretical Analysis. *Russ. J. Bioorg. Chem.* 2014; 4 (6): 612–9.

Eletto D, Dersh D, Argon Y. GRP94 in ER quality control and stress responses. *Semin. Cell. Dev. Biol.* 2010; 21 (5): 479-85.

Enomoto M, Kawazu S, Kawai S, Furuyama W, et al. Blockage of spontaneous Ca2+ oscillation causes cell death in intraerythrocitic *Plasmodium falciparum*. *PLoS One*. 2012; 7 (7): e39499.

Evans ML, Schmidt JC, Ilbert M, Doyle SM, et al. *E. coli* chaperones DnaK, Hsp33 and Spy inhibit bacterial functional amyloid assembly. *Prion*. 2012; 5 (4): 323-34.

Ferreira V, Molina MC, Valck C, Rojas A, et al. Role of calreticulin from parasites in its interaction with vertebrate hosts. *Mol. Imm.* 2004; 40: 1279–91.

Ferris SP, Jaber NS, Molinari M, Arvan P, Kaufman RJ. UDP-glucose: glycoprotein glucosyl transferase (UGGT1) promotes substrate solubility in the endoplasmic reticulum. *Mol Biol Cell*. 2013; 24 (17): 2597-608.

Field MC, Sergeenko T, Wang YN, Böhm S, Carrington M. Chaperone requirements for biosynthesis of the trypanosome variant surface glycoprotein. *PLoS One*. 2010; 5 (1): e8468.

Fletcher SM, Stark D, Harkness J, Ellis J. Enteric protozoa in the developed world: a public health perspective. *Clin. Microbiol. Rev*. 2012; 25: 420–49.

Florence L, Washburn M, Raine J, Anthony R, et al. A proteomic view of the *Plasmodium falciparum* life cycle. *Nature*. 2002; 419: 520-6.

Folgeira C, Carrión J, Moreno J, Saugar J, et al. Effects of the disruption of the HSP70-II gene on the growth, morphology, and virulence of *Leishmania intantum* promastigotes. *Int. Microbiol*. 2008; 11: 81-9.

Frölich S, Entzeroth R, Wallach M. 2011. Comparison of protective immune responses to apicomplexan parasites. *J. Parasitol. Res*. 2012; 2012: 852591.

Gammazza AM, Colangeli R, Orban G, Pierucci M, et al. Hsp60 response in experimental and human temporal lobe epilepsy. *Sci. Rep*. 2015; 5: 9434.

Georgescauld F, Popova K, Gupta AJ, Bracher A, et al. GroEL/ES chaperonin modulates the mechanism and accelerates the rate of TIM-barrel domain folding. *Cell*. 2014; 157: 922-34.

Gidalevitz T, Kikis EA, Morimoto RI. A cellular perspective on conformational disease: the role of genetic background and proteostasis networks. *Curr. Opin. Struct. Biol*. 2010; 20 (1): 23-32.

Gilberger T, Thompson J, Reed M, Good R, Cowman A. The cytoplasmic domain of the *Plasmodium falciparum* ligand EBA-175 is essential for invasion but not protein trafficking. *J. Cell. Biol*. 2003; 162 (2): 317-27.

Gitau G, Mandal P, Blatch G, Przyborski J. Characterization of the *Plasmodium falciparum* Hsp70-90 organizing protein (*Pf*Hop). *Cell. Stress Chaperones.* 2012; 17 (2):191-202.

González A, Valck C, Sánchez G, Härtel S, et al. *Trypanosoma cruzi* calreticulin topographical variations in parasites infecting murine macrophages. *Am. J. Trop. Med. Hyg.* 2015; 92 (5): 887-97.

Graefe S, Wiesgigl M, Gaworski I, Macdonald A, Clos J. Inhibition of HSP90 in *Trypanosoma cruzi* induces a stress response but no stage differentiation. *Eukaryot. Cell.* 2002; 1 (6): 936-43.

Griesemer M, Young C, Robinson AS, Petzold L. BiP clustering facilitates protein folding in the endoplasmic reticulum. *PLoS Comput. Biol.* 2014; 10 (7): e1003675.

Guevara J, Holder A, McBride J, Blackman M. Antibodies that inhibit malaria merozoite surface protein-1 processing and erythrocyte invasion are blocked by naturally acquired human antibodies. *J. Exp. Med.* 1997; 186 (1): 1689-99.

Guinn EJ, Jagannathan B, Marqusee S. Single-molecule chemo-mechanical unfolding reveals multiple transition state barriers in a small single-domain protein. *Nat. Commun.* 2015; 6: 6861.

Halliez MC and Buret AG. Extra-intestinal and long term consequences of *Giardia duodenalis* infections. *World J. Gastroenterol.* 2013; 19 (47): 8974–85.

Haque S, Majumdar T, Barik S. Redox-assisted protein folding systems in eukaryotic parasites. *Antioxid. Redox Signal.* 2012; 17 (4): 674-83.

Hartl FU, Hayer-Hartl M. Converging concepts of protein folding *in vitro* and *in vivo*. *Nat. Struct. Mol. Biol.* 2009; 16 (6): 574-81.

Hartley T, Siva M, Lai E, Teodoro T, et al. Endoplasmic reticulum stress response in an INS-1 pancreatic beta-cell line with inducible expression of a folding-deficient proinsulin. *BMC Cell. Biol.* 2010; 11: 59.

Hasin H, Cusack SA, Ali SS, Fitzpatrick DA, Jones GW. Global transcript and phenotypic analysis of yeast cells expressing Ssa1, Ssa2, Ssa3 or Ssa4 as sole source of cytosolic Hsp70-Ssa chaperone activity. *BMC Genomics.* 2014; 15: 194.

Hatherley R, Blatch GL, Bishop OT. *Plasmodium falciparum* Hsp70-x: a heat shock protein at the host-parasite interface. *J. Biomol. Struct. Dyn.* 2014; 32 (11): 1766-79.

Hecht O, Nuland N, Schleinkofer K, Dingley A, et al. solution structure of the pore-forming protein of *Entamoeba histolytica. J. Biol. Chem.* 2004; 279 (17): 17834-41.

Hehl A, Marti M, Köhler P. Stage-specific expression and targeting of cyst wall protein-green fluorescent protein chimeras in Giardia. *Mol. Biol. Cell.* 2000; 11:1789-1800.

Hodder A, Crewther P, Matthew M, Reid G, et al. The disulfide bond structure of Plasmodium apical membrane antigen-1. *J. Biol. Chem.* 1996; 271 (46): 29446-52.

Hong BX, Soong L. Identification and enzymatic activities of four protein disulfide isomerase (PDI) isoforms of *Leishmania amazonensis. Parasitol. Res.* 2008; 102 (3): 437-46.

Im C-N, Se J-S. Overexpression of tumor necrosis factor receptor-associated protein 1 (TRAP1), leads to mitochondrial aberrations in mouse fibroblast NIH/3T3 cells. *BMB Rep.* 2014; 47 (5): 280-5.

Jones C. Anderson S, Singha UK, Chaudhuri M. Protein phosphatase 5 is required for Hsp90 function during proteotoxic stresses in *Trypanosoma brucei. Parasitol. Res.* 2008; 102 (5): 835-44.

Joyce BR, Tampaki Z, Kim K, Wek RC, Sullivan WJ Jr. The unfolded protein response in the protozoan parasite *Toxoplasma gondii* features translational and transcriptional control. *Eukaryot. Cell.* 2013; 12 (7): 979-89.

Kasembeli M, Lau WCY, Roh S-H, Eckols TK, et al. Modulation of STAT3 folding and function by TRiC/CCT chaperonin. *PLoS Biol.* 2014; 12 (4): e1001844.

Kawabata S, Gills JJ, Mercado-Matos JR, Lopiccolo J, et al. Synergistic effects of nelfinavir and bortezomib on proteotoxic death of NSCLC and multiple myeloma cells. *Cell. Death Dis.* 2012; 3: e353.

Khalaf N, De Muylder G, Louzir H, McKerrow J, Chenik Mehdi. *Leishmania major* protein disulfide isomerase as a drug target: enzymatic and functional characterization. *Parasitol. Res.* 2012; 110 (5): 1911-7.

Kiktev DA, Melomed MM, Lu CD, Newnam GP, Chernoff YO. Feedback control of prion formation and propagation by the ribosome-associated chaperone complex. *Mol. Microbiol.* 2015; 96: 621–32.

Kim SJ, Yoon Js, Shishido H, Yang Z, et al. Translational tuning optimizes nascent protein folding in cells. *Science*. 2015; 348 (6233): 444-8.

Knodler L, Noiva R, Metha K, McCaffery J, et al. Novel protein-disulfide isomerases from the early-diverging protist *Giardia lamblia. J. Biol. Chem.* 1999; 274 (42): 29805-11.

Krobitsch S, Clos J. A novel role for 100 kD heat shock proteins in the parasite *Leishmania donovani. Cell. Stress Chaperones.* 1999; 4 (3): 191-8.

Kumar R, Pavithra SR, Tatu U. Three-dimensional structure of heat shock protein 90 from *Plasmodium falciparum*: molecular modelling approach to rational drug design against malaria. *J. Biosci.* 2007; 32 (3): 531-6.

Kumar R. Musiyenko A. Barik S. The heat shock protein 90 of *Plasmodium falciparum* and antimalarial activity of its inhibitor geldanamycin. *Malaria J.* 2003; 2: 30.

Lawen A. Biosynthesis of cyclosporins and other natural peptidyl prolyl cis/trans isomerase inhibitors. *Biochim. Biophys. Acta.* 2014. doi: 10.1016/j.bbagen.2014.12.009.

Li Q, Zhou Y, Yao C, Ma X, et al. Apoptosis caused by Hsp90 inhibitor geldanamycin in *Leishmania donovani* during promastigote-to-amastigote transformation stage. *Parasitol. Res.* 2009; 105 (6): 1539-48.

Li YP, Chen JJ, Shen JJ, Cui J, et al. Synthesis and biological evaluation of geldanamycin analogs against human cancer cells. *Cancer Chemother. Pharmacol.* 2015; (4): 773-82.

Lisanti S, Tavecchio M, Chae YC, Liu Q, et al. Deletion of the mitochondrial chaperone TRAP-1 uncovers global reprogramming of metabolic networks. *Cell. Reports.* 2014; 8: 671–7.

Liu B, Li Z. Endoplasmic reticulum HSP90b1 (gp96, grp94) optimizes B-cell function via chaperoning integrin and TLR but not immunoglobulin. Blood. 2008; 112 (4): 1223-30.

Liu L, Chen J, Yang B, Wang Y. Oligomer-dependent and -independent chaperone activity of sHsps in different stressed conditions. *FEBS Open Bio.* 2015; 5: 155–162.

Liu Y, Ye Y. Proteostasis regulation at the endoplasmic reticulum: a new perturbation site for targeted cancer therapy. *Cell Res.* 2011; 21 (6): 867-83.

Luo B, Lee AS. The critical roles of endoplasmic reticulum chaperones and unfolded protein response in tumorigenesis and anticancer therapies. *Oncogene.* 2013; 32 (7): 805-18.

Mahajan B, Noiva R, Yadava A, Zheng H, et al. Protein disulfide isomerase assisted protein folding in malaria parasites. *Int. J. Parasitol.* 2006; 36: 1037-48.

Mares RE, Magaña PD, Meléndez-López SG, Licea AF, et al. Oxidative folding and reductive activities of *Eh*PDI, a protein disulfide isomerase from *Entamoeba histolytica. Parasitol. Int.* 2009; 58 (3): 311-3.

Mares RE, Minchaca AZ, Villagrana S, Meléndez-López SG, Ramos MA. Analysis of the isomerase and chaperone-like activities of an amebic PDI (*Eh*PDI). *Biomed. Res. Int.* 2015; 2015: 286972.

Marín-Menéndez A, Bell A. Overexpression, purification and assessment of cyclosporin binding of a family of cyclophilins and cyclophilin-like proteins of the human malarial parasite *Plasmodium falciparum*. *Protein Expr. Purif.* 2011; 78 (2): 225-34.

Markikou-Ouni W, Achour-Chenik Y, Meddeb-Garnaoui A. Effects of *Leishmania major* clones showing different levels of virulence on infectivity, differentiation and maturation of human dendritic cells. *Clin. Exp. Immunol.* 2012; 169: 273-80.

Marzec M, Eletto D, Argon Y. GRP94: An HSP90-like protein specialized for protein folding and quality control in the endoplasmic reticulum. *Biochim. Biophys. Acta.* 2012; 1823 (3): 774-87.

Meyer KJ, Shapiro TA. Potent antitrypanosomal activities of heat shock protein 90 inhibitors in vitro and in vivo. *J Infect Dis*. 2013; 208 (3): 489-99.

Michaeli S. The response of trypanosomes and other eukaryotes to ER stress and the spliced leader RNA silencing (SLS) pathway in *Trypanosoma brucei*. *Crit. Rev. Biochem. Mol. Biol.* 2015: 1-12.

Miller M, McGowan S, Grantt K, Champion M, et al. Inducible resistance to oxidant stress in the protozoan *Leishmania chagasi*. *J. Biol. Chem.* 2000; 272 (43): 33883-9.

Moss D, Remarque E, Faber B, Cavanagh D, et al. *Plasmodium falciparum* 19-kilodalton merozoite surface protein 1 (MSP1)-specific antibodies that interfere with parasite growth in vitro can inhibit MSP1 processing, merozoite invasion, and intracellular parasite development. *Infect. Immun.* 2012; 80 (3): 1280-7.

Mout R, Xu ZD, Wolf AK, Jo Davisson V, Jarori GK. Anti-malarial activity of geldanamycin derivatives in mice infected with *Plasmodium yoelii*. *Malar. J.* 2012; 11: 54.

Nagajyothi F, Machado F, Burleigh A, Jelicks L, et al. Mechanisms of *Trypanosome cruzi* persistence in Chagas disease. *Cell. Microbiol.* 2012; 14 (5): 634-43.

Nageshan R, Roy N, Hehl A, Tatu U. Post-transcriptional repair of a split heat shock protein 90 gene by mRNA trans-splicing. *J. Biol. Chem.* 2011; 286 (9): 7116-22.

Neckers L, Tatu U. Molecular chaperones in pathogen virulence: emerging new targets for therapy. *Cell Host Microbe*. 2008; 4 (6): 519-27.

Ostoa-Saloma P, Carrero JC, Petrossian P, Herión P, *et al.* Cloning, characterization and functional expression of a cyclophilin of *Entamoeba histolytica*. *Mol. Biochem. Parasitol.* 2000; 107 (2): 219-25.

Paba J, Ricart C, Fontes W, Santana J, et al. Proteomic analysis of *Trypanosoma cruzi* developmental stages using isotope-coded affinity tag reagents. *J. Proteome Res.* 2004; 3 (3): 517-24.

Pallavi R, Roy N, Nageshan RK, Talukdar P, et al. Heat shock protein 90 as a drug target against protozoan infections: biochemical characterization of HSP90 from *Plasmodium falciparum* and *Trypanosoma evansi* and evaluation of its inhibitor as a candidate drug. *J Biol Chem.* 2010; 285 (49): 37964-75.

Pan S, Chen HH, Correia C, Dai H, et al. Cell surface protein disulfide isomerase regulates natriuretic peptide generation of cyclic guanosine monophosphate. *PLoS ONE.* 2014; 9 (11): e112986.

Papa FR. Endoplasmic reticulum stress, pancreatic beta-cell degeneration, and diabetes. *Cold Spring Harb. Perspect. Med.* 2012; 2 (9): a007666.

Peters LR, Raghavan M. Endoplasmic reticulum calcium depletion impacts chaperone secretion, innate immunity, and phagocytic uptake of cells. *J. Immunol.* 2011; 187 (2): 919-31.

Petersen AL, Guedes CE, Versoza CL, Lima JG, et al. 17-AAG kills intracellular *Leishmania amazonensis* while reducing inflammatory responses in infected macrophages. *PLoS One*. 2012; 7 (11): e49496.

Raina P, Kaur S. Knockdown of LdMC1 and Hsp70 by antisense oligonucleotides causes cell-cycle defects and programmed cell death in *Leishmania donovani. Mol. Cell. Biochem.* 2012; 359: 135-49.

Ralston KS, Petri Jr WA. 2011. Tissue destruction and invasion by *Entamoeba histolytica*. Review. *Trends Parasitol.* 2011; 27 (6): 254-63.

Ramirez C, Dea-Ayuela M, Gutiérrez-Blázquez M, Bolas-Fernández F, et al. Identification of ptoreins interacting with HSP70 mRNAs in *Leishmania brazilensis*. *J. Proteomics*. 2013; 94: 124-37.

Ramos M, Mares R, Magaña P, Ortega J, Cornejo-Bravo J. *In silico* identification of the protein disulfide isomerase family from a protozoan parasite. *Comput. Biol. Chem.* 2008; 32 (1): 66-70.

Ramos MA, Mares RE, Magaña PD, Rivas ID, Meléndez-López SG. *Entamoeba histolytica*: biochemical characterization of a protein disulfide isomerase. *Exp. Parasitol.* 2011; 128 (1): 76-81.

Ramos MA, Sanchez-Lopez R, Mares RE, Olvera F, Alagón A. Identification of an *Entamoeba histolytica* gene encoding a protein disulfide isomerase that functionally complements the *dsb*A mutation in *Escherichia coli*. *Mol. Biochem. Parasitol.* 2005; 143 (2): 236-40.

Ratner DM, Cui J, Steffen M, Moore LL, *et al*. Changes in the N-glycome, glycoproteins with Asn-linked glycans, of *Giardia lamblia* with differentiation from trophozoites to cysts. *Eukaryot. Cell.* 2008; 7 (11): 1930-40.

Reidy M, Sharma R, Shastry S, Roberts B-L, et al. Hsp40s specify functions of Hsp104 and Hsp90 protein chaperone machines. *PLoS Genet.* 2014; 10 (10): e1004720.

Rosenberger K, Dembny P, Derkow K, Engel O, et al. Intrathecal heat shock protein 60 mediates neurodegeneration and demyelination in the CNS through a TLR4- and MyD88-dependent pathway. *Mol. Neurodegener.* 2015; 10: 5.

Rubotham J, Woods K, Garcia-Salcedo J, Pays E, Nolan D. Characterization of two protein disulfide isomerases from the endocytic pathway of bloodstream forms of *Trypanosoma brucei*. *J. Biol. Chem.* 2005; 280 (11): 10410-8.

Rüßmann F, Stemp MJ, Mönkemeyer L, Etchells SA, et al. Folding of large multidomain proteins by partial encapsulation in the chaperonin TRiC/CCT. *Proc. Natl. Acad. Sci. U.S.A.* 2012; 109 (52): 21208-15.

Saibil H. Chaperone machines for protein folding, unfolding and disaggregation. *Nat. Rev. Mol. Cell Biol.* 2013; 14 (10): 630–42.

Samuelson J, Robbins P. A simple fibril and lectin model for cyst walls of Entamoeba and perhaps Giardia. *Trends Parasitol.* 2011; 27 (1): 17-22.

Santos C, Stolf B, Takemoto P, Amanso A, et al. Protein dusulfide isomerase (PDI) associates with NADPH oxidase and is required for phagocytosis of *Leishmania chagasi* promastigotes by macrophages. *J. Leukocyte Biol.* 2009; 86 (4): 989-98.

Santos DM, Petersen AL, Celes FS, Borges VM, et al. Chemotherapeutic potential of 17-AAG against cutaneous leishmaniasis caused by *Leishmania (Viannia) braziliensis*. *PLoS Negl. Trop Dis.* 2014; 8 (10):e3275.

Schiene-Fischer C. Multidomain peptidyl prolyl cis/trans isomerases. *Biochim. Biophys. Acta.* 2014. doi: 10.1016/j.bbagen.2014.11.012.

Schmidpeter PA, Schmid FX. Prolyl isomerization and its catalysis in protein folding and protein function. *J. Mol. Biol.* 2015; 427 (7): 1609-31.

Shahinas D, Folefoc A, Taldone T, Chiosis G, et al. A purine analog synergizes with chloroquine (CQ) by targeting *Plasmodium falciparum* Hsp90 (*Pf*Hsp90). *PLoS One.* 2013; 8 (9): e75446.

Shahinas D, Macmullin G, Benedict C, Crandall I, Pillai DR. Harmine is a potent antimalarial targeting Hsp90 and synergizes with chloroquine and artemisinin. *Antimicrob. Agents Chemother.* 2012; 56 (8): 4207-13.

Shaner L, Wegele H, Buchner J, Morano KA. The Yeast Hsp110 Sse1 functionally interacts with the Hsp70 chaperones Ssa and Ssb. *J. Biol. Chem.* 2005; 280 (50): 41262–9.

Shonhai A, Boshoff A, Blatch G. The structural and functional diversity of Hsp70 proteins from *Plasmodium falciparum*. *Prot. Sci.* 2007; 16: 1803-18.

Sibley LD. The roles of intramembrane proteases in protozoan parasites. *Biochim. Biophys. Acta.* 2013; 1828 (12): 2908-15.

Singh M, Shah V, Tatu U. A novel C-Terminal homologue of aha1 co-chaperone binds to heat shock protein 90 and stimulates its ATPase activity in *Entamoeba histolytica*. *J. Mol. Biol.* 2014; 426 (8): 1786-98.

Staron M, Yang Y, Liu B, Li J, *et al.* gp96, an endoplasmic reticulum master chaperone for integrins and Toll-like receptors, selectively regulates early T and B lymphopoiesis. *Blood.* 2010; 115 (12): 2380-90.

Stolf BS, Smyrnias I, Lopes LR, Vendramin A, et al. Protein disulfide isomerase and host-pathogen interaction. *Scientific World Journal.* 2011; 11: 1749-61.

Tannous A, Pisoni GB, Hebert DN, Molinari M. N-linked sugar-regulated protein folding and quality control in the ER. *Semin. Cell. Dev. Biol.* 2015. doi: 10.1016/j.semcdb.2014.12.001.

Teixeira F, Castro H, Cruz T, Tse E, et al. Mitochondrial peroxiredoxin functions as crucial chaperone reservoir in *Leishmania infantum*. *Proc. Natl. Acad. Sci. U.S.A.* 2015; 112 (7): E616-24.

Wang YN, Wang M, Field MC. *Trypanosoma brucei*: trypanosome-specific endoplasmic reticulum proteins involved in variant surface glycoprotein expression. *Exp. Parasitol.* 2010; 125 (3): 208-21.

WHO, World Health Organization. April 2015. Malaria Fact Sheet N°94. http://www.who.int/mediacentre/factsheets/fs094/en/ (Accessed 06.25.2015).

WHO, World Health Organization. February 2015. Leishmaniasis Fact sheet N°375. http://www.who.int/mediacentre/factsheets/fs375/en/ (Accessed 06.25.2015).

WHO, World Health Organization. March 2015. Chagas Disease Fact sheet N°340. http://www.who.int/mediacentre/factsheets/fs340/en/ (Accessed 06.25.2015).

WHO, World Health Organization. World Health Statistics 2014. http://www.who.int/gho/publications/world_health_statistics/2014/en/ (Accessed 06.25.2015).

WHO/PAHO/UNESCO report. A consultation with experts on amoebiasis. *Epidemiol. Bull.* 1997; 18: 13–14.

Wiesgigl M, Clos J. Heat shock protein 90 homeostasis controls stage differentiation in *Leishmania donovani*. *Mol. Biol. Cell.* 2001; 12 (11): 3307-16.

Ximénez C, González E, Nieves ME, Silva-Olivares A, et al. *Entamoeba histolytica* and *E. dispar* calreticulin: inhibition of classical complement pathway and differences in the level of expression in amoebic liver abscess. *Biomed. Res. Int.* 2014; 2014: 127453.

Xu C, Bailly-Maitre B, Reed JC. Endoplasmic reticulum stress: cell life and death decisions. *J. Clin. Invest.* 2005; 115 (10): 2656-64.

Yau WL, Blisnick T, Taly JF, Helmer-Citterich M, et al. Cyclosporin A treatment of *Leishmania donovani* reveals stage-specific functions of cyclophilins in parasite proliferation and viability. *PLoS Negl. Trop. Dis.* 2010; 4 (6): e729.

Yau WL, Pescher P, MacDonald A, Hem S, et al. The *Leishmania donovani* chaperone cyclophilin 40 is essential for intracellular infection independent of its stage-specific phosphorylation status. *Mol. Microbiol.* 2014; 93 (1): 80-97.

Zhang K, Ezemaduka AN, Wang Z, Hu H, et al. A Novel mechanism for small heat shock proteins to function as molecular chaperones. *Sci. Rep.* 2015; 5: 8811.

Zhu G, Lee AS. Role of the unfolded protein response, GRP78 and GRP94 in organ homeostasis. *J. Cell. Physiol.* 2015; 230: 1413–20.

In: New Research on Molecular Chaperones ISBN: 978-1-63483-543-5
Editor: Jaime Wyatt 

*Chapter 2*

# PDIA1/P4HB AS MODEL OF ENZYME WITH MOLECULAR CHAPERONE FUNCTION

***Rosa E. Mares, Alexis Z. Minchaca, Patricia L. A. Muñoz and Marco A. Ramos****

Chemical Sciences and Engineering Faculty,
Autonomous University of Baja California, Tijuana, Mexico

## ABSTRACT

Protein folding is a sensitive process that can be affected by several factors, among which pH and temperature are notable (environmental factors). Protein misfolding is an important issue in some human disorders, e.g., Alzheimer's disease. Molecular chaperones are multi-domain proteins with a specific function in folding, unfolding, and remodeling of proteins. *In vivo*, molecular chaperones assist nascent polypeptides throughout the folding process until the native conformation has been attained. Furthermore, they can regulate unfolding and aggregation, protecting the cell from a proteotoxic stress. Protein disulfide isomerase (PDI, EC 5.3.4.1) enzymes are oxidoreductases that catalyze the oxidation and rearrangement of disulfide bonds in polypeptide substrates. PDI enzymes share as a common feature an active domain that folds similar to the bacterial protein thioredoxin (thus named thioredoxin domain), which has the conserved CXXC motif at its active

* To whom should correspondence be addressed: Marco A. Ramos, PhD., E-mail: mramos@uabc.edu.mx.

site. The protein disulfide isomerase A1 (PDIA1), also named beta subunit of prolyl-4-hydroxylase (P4HB), is a multifunctional enzyme that plays a key role in the dithiol-disulfide equilibrium of the secretory pathway. Interestingly, a chaperone-like activity has been observed at high concentration, inhibiting aggregation of misfolded proteins. Conversely, anti-chaperone activity has been noted at low concentrations, promoting aggregation. This chapter reviews the current progress regarding the structure-function relationship as well as its association with the chaperone-like activity of the human PDIA1/P4HB protein.

# 1. INTRODUCTION

The native tridimensional structure of proteins is ensured by the cellular process known as protein folding, which is conducted by the functional activity of two important assistants: folding enzymes (foldases) and molecular chaperones. Foldases reduce the reaction time of slow biochemical processes, such as disulfide bond formation or proline isomerization. While molecular chaperones bind to unfolded polypeptides, preventing aggregation and favoring correct folding, without both chemically-modifying the polypeptide substrate and affecting its folding rate.

To protect the integrity of their proteome, cells have an extensive regulatory network aimed to maintain proteostasis (defined as the cellular state where prevails a dynamic equilibrium between synthesis, folding, and degradation of proteins), which allows sufficient flexibility to get a healthy cellular function. This proteostatic network (PN) comprises a set of foldases and chaperones as well as a mechanism of proteolytic degradation. During sustained stress conditions, PN declines in performance, producing as a final outcome the accumulation of cytotoxic protein aggregates, which are associated with several human disorders, such as neurodegenerative diseases.

The protein disulfide isomerases (PDI) are ER-resident oxidoreductases that catalyze the formation and rearrangement of disulfide bonds in nascent polypeptides. Also, they can prevent the aggregation of unfolded proteins, functioning as molecular chaperone. Structurally, PDI are multi-domain proteins that comprise at least one thioredoxin-like domain. Redox active enzymes are characterized by the presence of a functional domain (a-type) that contains the conserved catalytic motif CXXC on its active site.

The human PDI family comprises 21 members, varying in size, expression, localization, and enzymatic function. The protein disulfide isomerase A1 (PDIA1), also named beta subunit of prolyl-4-hydroxylase

(P4HB), is a multifunctional enzyme that plays a key role in the dithiol-disulfide equilibrium of the secretory pathway. PDIA1 is mainly located within the ER, where displays many versatile functions in maintaining protein homeostasis. However, it has also been found in other cellular locations, such as the cell surface, where it has been associated with cell adhesion and pathogen internalization.

Numerous studies have shown experimental and clinical data suggesting consistent association between defective production and function of PDIA1 with development of human pathologies. Consequently, its application as a potential drug target has attracted increasing attention. However, before clinical trials, some safety issues should be addressed to demonstrate the feasibility of targeting PDIA1.

## 2. Cellular Proteostasis

In cells, proteins participate in all biological processes. Therefore, they can be regarded as the most functionally important biomolecules. In addition to their amino acid sequence, other protein properties, such as the level of expression, specific activity, post-translation modifications, and association with other biomolecules, are essential to function properly (Aebersold 2003). Hence, to protect the integrity of their proteome, cells have an extensive regulatory network aimed to maintain proteostasis (defined as the cellular state where prevails a dynamic equilibrium between synthesis, folding, and degradation of proteins), which allows sufficient flexibility to get a healthy cell function (Gidalevitz 2010; Hipp 2014). Interestingly, this proteostatic network (PN) includes a set of molecular chaperones and foldases, as well as a mechanism of proteolytic degradation (Kim 2013). During sustained stress conditions, PN declines in performance and nascent polypeptides are not able to fold efficiently, producing metastable proteins with functionally inactive conformations. Furthermore, a critical outcome of PN reduction is the accumulation of cytotoxic protein aggregates, which are associated with several neurodegenerative diseases and other medically important disorders (Hipp 2014).

From a physiological perspective, the entire process for functional proteins includes transcription, translation, trafficking and processing, assembly, localization, and degradation (Mu 2008). As they promote a healthy cell physiology, loss of function contributes to toxic effects that may result in cellular phenotypes associated with diseases (Gidalevitz 2010). Furthermore, during cellular aging, the accumulation of misfolded proteins is a constant challenge for PN, which eventually leads to the gathering of protein aggregates. The inability of cells to restore proteostasis may result in illness, and even death (Hartl 2011).

# 3. ER Stress and Unfolded Protein Response

ER is an important organelle involved in folding and post-translational modification of numerous proteins. As this subcellular compartment stablishes a close association with other organelles, such as mitochondria, Golgi complex, and nucleus, it is regarded as a central player in cellular physiology. Since there is a large production of proteins in the ER, preventing the misfolding of proteins it is important for the organelle. Moreover, protein Folding capacity of the ER is adjusted according to cell requirements in order to maintain proteostasis (Walker 2011). The accumulation of misfolded proteins in ER elicits a cellular response including: (1) reduction or inhibition in the rate of protein synthesis, (2) up-regulation of genes encoding ER chaperones, enzymes, and structural components of the ER, allowing it to process more proteins, and (3) initiation of apoptosis if stress is too severe (Sanderson 2015).

## 3.1. ER Stress

There are several physiological and pathological conditions that may lead to ER stress, such as nutrient deprivation, hypoxia, decreased ER $Ca^{2+}$ concentration, reduced glycosylation or disulfide bond formation, oxidative stress, and viral or bacterial infections, among others. ER stress occurs when the protein load exceeds its capacity to fold or degrade them, and is manifested by the accumulation of misfolded proteins. When this occurs, a highly conserved mechanism, the unfolded protein response (UPR), which aims at restoring ER homeostasis by activating a signaling cascade to transiently reduce translation, to induce molecular chaperones and foldases that enhance the capacity of protein folding, and to initiate a process of export and degrade misfolded proteins (Luo 2013).

To support proper folding and prevent aggregation in the ER, numerous resident chaperones and foldases assist protein maturation by signal peptide cleavage, glycosylation, and disulfide bond formation. In particular, chaperones participate in every aspect of the ER quality control. The most abundant and best characterized is the chaperone BiP/Grp78 (immunoglobulin heavy chain binding protein or glucose regulated protein 78), an Hsp70 family ATPase involved in several functions, such as translocation of nascent polypeptides, facilitating *de* novo folding and assembly, targeting misfolded proteins to ERAD (ER-associated degradation), and maintaining calcium homeostasis. To maintain protein folding homeostasis in the ER, the cell must balance the folding load with ample folding machinery, particularly chaperones such as BiP (Gardner 2013).

## 3.2. Unfolded Protein Response

To deal with ER stress and maintain homeostasis, cells activate the UPR mechanism, which coordinates the increase in protein folding capacity through a transcriptional up-regulation of folding, lipid synthesis, and ERAD machinery with a decrease in folding load through selective mRNA degradation and translational repression. UPR is a cytoprotective mechanism that allows cell adaptation to developmental and environmental conditions that affect ER folding. However, during severe and prolonged ER stress, UPR becomes cytotoxic and induces apoptosis. This outcome is an important pathogenic factor in a number of diseases, including diabetes, neurodegenerative disorders, atherosclerosis, and renal disease. Interestingly, because UPR plays a central role in determining cell fate, numerous studies have been conducted to identify modulators of the mechanism and exploit UPR for therapeutic benefit (Gardner 2013).

## 3.3. Homeostatic Mechanism of the UPR

The UPR is a homeostatic mechanism mediated by the activation of three upstream ER stress sensors: inositol-Requiring kinase 1 (IRE1), PKR-like endoplasmic reticulum kinase (PERK), and activating transcription factor 6 (ATF6) (Walker 2011).

IRE1 is a conserved transmembrane protein with an ER-luminal domain that senses misfolded proteins in the ER, mostly by direct ligand recognition. As consequence, IRE1 activates its cytoplasmic kinase and endoribonuclease domains, which initiate a non-conventional mRNA splicing reaction that results in the cleavage of the mRNA encoding the transcription factor XBP1

(X-box binding protein 1), the most important effector of the UPR that triggers the transcription of a plethora of ER stress responsive genes, including chaperones and foldases (Li 2010). Moreover, during severe ER stress conditions, IRE1 recruits TNF receptor-associated factor 2 and apoptosis signal-regulating kinase 1, then activates c-Jun N-terminal kinase (JNK) and induces apoptosis. Also, it has been showed that IRE1 splices certain microRNAs that inhibit caspase 2 expression (Chen 2014). The IRE1 pathway is the most conserved branch of the UPR. For instance, mammalian cells encode two IRE1 isoforms: IRE1α and IRE1β. The α-isoform is the most abundant and is expressed in almost all cells and tissues, while the β-isoform is expressed to significant levels only in intestinal epithelial cells (Stahl 2013).

PERK is a transmembrane Ser/Thr protein kinase that contains an N-terminal ER -luminal domain and a C-terminal cytoplasmic protein kinase domain. PERK phosphorylates two substrates: the eukaryotic translation initiation factor 2 alpha (eIF2α) and the Nrf2 transcription factor. Phosphorylation of eIF2α attenuates translation initiation of most transcripts while concurrently increases translation of select mRNAs (Bobrovnikova-Marjon 2010). Thereby, reducing polypeptide synthesis and, thus, the load of proteins entering into the ER (Lin 2009). PERK-dependent attenuation of protein synthesis contributes to redox homeostasis by decreasing consumption of reducing equivalents necessary for the formation of disulfide bonds during oxidative folding in the ER. In addition, PERK-dependent phosphorylation of Nrf2 releases it from an inhibitory complex, allowing the accumulation and nuclear import of Nrf2, favoring the induction of enzymes critical for the elimination of intracellular reactive oxygen species (ROS). Thus, PERK function is critical for maintaining cellular redox homeostasis and reducing ROS-induced genotoxic stress (Bobrovnikova-Marjon 2010).

ATF6 is a transmembrane protein that is activated by translocation from ER to Golgi and by proteolysis. Following cleavage, ATF6-p50 (the cytosolic DNA binding fragment) is released and translocated into the nucleus, where it binds to ER stress elements (ERSE) to amplify the transcription of UPR genes. Also, ATF6 targets processed XBP1 to increase its binding to ERSE and enhance transcription of genes needed for the UPR mechanism (Sanderson 2015).

The integrated responses elicited by the three ER stress sensors contribute to either adaptation or death, but it remains unknown how each pathway impacts to the ultimate biological effect. Numerous studies have attempted to solve this issue. For instance, IRE1 and ATF6 activities are attenuated by persistent ER stress, while PERK activity is not. Indeed, sustained PERK signaling impairs cell proliferation and promotes apoptosis. Conversely, signaling through IRE1 for an equivalent period of time increases cell proliferation without promoting death, suggesting that differential activation of PERK and IRE1 may determine cell fate after ER stress (Bravo 2013).

## 4. HUMAN PDI FAMILY

Mammalian cells produce a vast quantity of proteins with complicated disulfide bonded structures (e.g., nearly one third of total human proteins that transit the secretory pathway). The synthesis and folding of these type of proteins is carried out in the ER compartment, where the environment is favorable for disulfide bond formation, e.g., the ER-lumen is more oxidizing than the cytosol and encloses the enzymes required for catalyzing the biochemical reaction (Appenzeller-Herzog 2008).

Protein disulfide isomerases (PDI) are abundant ER-resident oxidoreductases catalyzing the formation and rearrangement of disulfide bonds in nascent polypeptides. Also, they can prevent the aggregation of unfolded proteins, functioning as molecular chaperones (Appenzeller-Herzog 2008; Araki 2013). Structurally, PDI enzymes are multi-domain proteins that may contain one or more thioredoxin-like domains in different arrangements. The catalytic motif CXXC, responsible for the oxidoreductase activity, is contained in the active domains (*a*-type). Whilst redox inactive domains (*b*-type), because they lack the catalytic motif, play important roles in substrate binding.

The first three-dimensional structure solved for a human PDI was the N-terminal thioredoxin-like *a*-type domain of PDIA1. The structural features of this clearly improve the understanding on how PDI enzymes function: (1) the secondary structure elements are arranged in a β1-α1-β2-α2-β3-α3-β4-β5-α4 fashion, (2) the catalytic motif residues are positioned at the N-terminus of the second α-helix, and (3) the first catalytic Cys residue is partially exposed to the surface. Also, a hydrophobic patch and a loop containing a Pro residue in the cis-conformation were found in the vicinity of the active site, both implicated in substrate binding (Appenzeller-Herzog 2008; Wang 2012).

The human PDI family comprises 21 members (Table 1), varying in size, expression, localization, and enzymatic function. Although the family name implies a disulfide isomerase function, this activity has not been experimentally demonstrated for all members. However, a unifying feature is the presence of at least one thioredoxin-like domain, which can be either catalytically active or inactive (Appenzeller-Herzog 2008; Galligan 2012).

**Table 1. Human PDI family members as listed in the Human Gene Nomenclature Committee Database (http://www.genenames.org/cgi-bin/genefamilies/set/692)**

| Approved Symbol | Approved Name | Previous Symbols | Synonyms |
|---|---|---|---|
| P4HB | prolyl 4-hydroxylase, beta polypeptide | PO4DB, ERBA2L | GIT, DSI, P4Hbeta, PO4HB, PDIA1, PDI, PROHB |
| PDIA2 | protein disulfide isomerase family A, member 2 | PDIp | PDA2, PDI, PDIR |
| PDIA3 | protein disulfide isomerase family A, member 3 | GRP58 | P58, ERp61, ERp57, ERp60, GRP57, PI-PLC, HsT17083 |
| PDIA4 | protein disulfide isomerase family A, member 4 | | ERP70, ERP72 |
| PDIA5 | protein disulfide isomerase family A, member 5 | | PDIR, FLJ30401 |
| PDIA6 | protein disulfide isomerase family A, member 6 | TXNDC7 | P5, ERp5 |
| PDILT | protein disulfide isomerase-like, testis expressed | | PDIA7 |
| ERP27 | endoplasmic reticulum protein 27 | C12orf46 | ERp27, FLJ32115, PDIA8 |
| ERP29 | endoplasmic reticulum protein 29 | C12orf8 | PDIA9, PDI-DB, ERp29, ERp31, ERp28 |
| ERP44 | endoplasmic reticulum protein 44 | TXNDC4 | KIAA0573, PDIA10 |
| TMX1 | thioredoxin-related transmembrane protein 1 | TXNDC, TXNDC1 | TMX, PDIA11 |
| TMX2 | thioredoxin-related transmembrane protein 2 | TXNDC1 4 | PDIA12 |
| TMX3 | thioredoxin-related transmembrane protein 3 | TXNDC1 0 | FLJ20793, PDIA13, KIAA1830 |
| TMX4 | thioredoxin-related transmembrane protein 4 | TXNDC1 3 | PDIA14, DJ971N18.2, KIAA1162 |
| TXNDC5 | thioredoxin domain containing 5 (endoplasmic reticulum) | | ERp46, Hcc-2, PDIA15, FLJ21353, EndoPDI, FLJ90810, MGC3178 |
| TXNDC12 | thioredoxin domain containing 12 (endoplasmic reticulum) | | PDIA16, AGR1, TLP19, hAG-1, ERP18, ERP19 |
| AGR2 | anterior gradient 2, protein disulphide isomerase family member | | PDIA17, HAG-2, AG2, XAG-2 |
| AGR3 | anterior gradient 3, protein disulphide isomerase family member | | PDIA18, HAG3, hAG-3, BCMP11 |
| DNAJC10 | DnaJ (Hsp40) homolog, subfamily C, member 10 | | ERdj5, PDIA19 |
| CASQ1 | calsequestrin 1 (fast-twitch, skeletal muscle) | CASQ | PDIB1 |
| CASQ2 | calsequestrin 2 (cardiac muscle) | | PDIB2 |

Detailed information on structural and functional features of each member can be obtained elsewhere (Appenzeller-Herzog 2008; Hatahet 2009; Coe 2010; Kozlov 2010; Imaoka 2011; Stolf 2011; Benham 2012; Laurindo 2012; Shishkin 2013; Lu 2014).

# 5. PDIA1/P4HB As Foldase with Chaperone-Like Activity

PDIA1, also named P4HB and traditionally referred as PDI, is an ER-resident foldase catalyzing the oxidation, reduction, and isomerization of disulfide bonds in substrate proteins. Also, it may exhibit an ATP-independent chaperone-like activity, preventing misfolding and aggregation of proteins (Cai 1994; Ali Khan 2011). In addition to these, it has been identified as the β-subunit of prolyl-4-hydroxylase and as a component of the microsomal triglyceride transfer complex (Galligan 2012). This foldase consists of 508 amino acid residues, including a 17-residue N-terminal signal peptide, two active domains (*a* and *a′*), both with the catalytic motif CGHC, two inactive domains (*b* and *b′*), and the C-terminal KDEL sequence as the ER-retention signal (Pihlajaniemi 1987).

The structure of PDIA1 revealed that the four thioredoxin-like domains (*a*-*b*-*b′*-*a′*) are arranged as a horseshoe shape, where domains *a* and *a′* are facing each other and exposed to the solvent, allowing interaction with their respective substrates, and domains *b* and *b′* are forming a rigid hydrophobic base, providing a substrate binding site. In addition, domains *a′* and *b′* are connected by a 19-residue inter-domain link, named x-linker, which promotes flexibility for substrate specificity (Wang 2010; Wang 2013). Furthermore, the protein folding pattern showed different conformations in the reduced and oxidized state, where the distance between the two active site regions increases from 27.6 to 40.3 Å, offering an open structure with more exposed hydrophobic areas when is oxidized, supporting a role for redox in chaperone-like activity (Wang 2013).

Although the catalytic activity of PDIA1 has been widely analyzed, the chaperone-like function has been slightly less studied. The initial efforts to get insights into this function showed that domain *b′* is the essential core of the substrate binding site and is sufficient to bind small peptides (Klappa 1998). However, participation of further domains provides additional binding interactions that are essential to bind large and more complex substrates (Klappa 1998; Wang 2012).

Recently, diminished chaperone-like activity of PDIA1 was noted as an outcome of redox-regulated inter-domain conformational changes, located within domains *b′* and *a′*, and triggered by redox switch of the active site in domain *a′*. Moreover, the crystal structure of the reduced state revealed a novel domain conformation: domain *a′* interacts with both the domain *b′* and the x-linker, forming a tightly compact module. Interestingly, upon oxidation of domain *a′*, that module relaxes exposing the hydrophobic regions of domain *b′*. Hence, it is likely to assume that such conformational changes induced by different redox states may lead to potential regulation of chaperone-like activity of PDIA1 (Wang 2012; Wang 2013).

## 6. Role of PDIA1/P4HB in Human Diseases

As a multi-functional ER-resident protein, PDIA1 behaves as oxidoreductase with chaperone-like activity, facilitates the ER-associated degradation of proteins, and participates as an essential component of active complexes. Additionally, it has been found in non-ER locations, e.g., the cytoplasmic membrane, where catalyzes the reduction of disulfide bonds of surface proteins, enhancing cell adhesion and migration (Pan 2014). Therefore, this protein is thought essential in maintaining cells and tissues in a healthy state.

Recently, PDIA1 has been associated with the development of diverse human diseases. For instance, as a key player for adhesion and internalization of some pathogens, as a promoter for the increment of ROS concentration in cells, and as a critical participant in the control of neurodegenerative diseases related to protein misfolding and aggregation (Wang 2012; Ali Khan 2014). As complex protein with important roles in both physiology and pathology, further studies are needed to gain insightful knowledge regarding the precise function of PDIA1 in health and disease (Ali Khan 2014).

### 6.1. Viral infections

An increasing number of studies supporting an important role for thiol-disulfide exchange during the entry of several viruses into susceptible cells, such as Sindbis virus, Baculovirus, Vaccina virus, Equine arteritis virus,

Moloney murine leukemia virus, Newcastle disease virus, Hepatitis delta virus, and HIV (Stantchev 2012). The role of PDIA1 on the initial stages of a viral infection is well-documented for HIV: the enzyme located on the cell surface of lymphocytes catalyzes the rearrangement of disulfide bonds in the viral envelope glycoproteins that promote virus-membrane fusion and viral entry (Bi 2011; Cerutti 2014). Furthermore, the observed blockage of HIV infection by PDI inhibitors revealed the importance of thiol-disulfide exchange and that PDIA1 is a key enzyme involved in the HIV entry process (Stantchev 2012).

## 6.2. Cancer and other proliferative diseases

High levels of PDIA1 have been identified in numerous malignant proliferative processes in humans, such as ovarian, prostate, and lung cancer, as well as lymphoma, glioma, leukemia, and melanoma (Xu 2012; Xu 2014). Conversely, low levels of PDIA1 have been associated with a higher survival rate in patients with certain cancer, suggesting a role in survival of malignant cells. Hence, suppression of apoptosis by PDIA1 has been proposed as a mechanism for tumor growth and metastasis (Parakh 2015). Moreover, inhibition of its activity leads to apoptosis in cancer, suggesting that PDIA1 is a promising target for drug development (Xu 2012).

## 6.3. Neurodegenerative diseases

Neurodegenerative diseases (ND) are characterized by their property to accumulate, insoluble and aggregated, misfolded proteins within the affected tissues. Although differing in the specific group of neurons affected, most common disorders, such as Alzheimer´s disease (AD), Parkinson´s disease (PD), Huntington´s disease (HD), amyotrophic lateral sclerosis (ALS), and Creutzfeldt-Jakob disease (CJD), differ in the type of misfolded protein that aggregates. For instance, β-amyloid protein in AD, α-synuclein protein in PD, and huntingtin protein in HD (Halloran 2013). ER stress induced by protein misfolding leads to up-regulation of several chaperones and foldases. Interestingly, PDIA1 is up-regulated in some ND, such as AD, CJD, and ALS (Halloran 2013). Furthermore, *in vitro* studies revealed that PDIA1 effectively inhibits α-synuclein fibril formation (Chen 2010), suggesting a relevant role in protecting cells to the effects of misfolded protein accumulation. Therefore, as part of a cellular protective mechanism, PDIA1 prevents protein misfolding and aggregation in neurodegeneration, by reducing the load of protein aggregates, and thus increasing neuronal viability (Grek 2014; Parakh 2015).

## 6.4. Metabolic diseases

The specialized function of pancreatic β-cells to produce and secrete insulin in response to changes in blood glucose concentrations has made them highly dependent on a remarkably efficient UPR. Diabetes mellitus (DM), a metabolic disease, is characterized by chronic high blood glucose levels. The two major types of DM are categorized based on insulin production: absolutely deficient (type 1) and relatively deficient (type 2) (Fonseca 2009). Resistance to insulin action in peripheral tissues, e.g., muscle and liver, is one of the primary features of DM type 2. This resistance leads to hyper-production of insulin in pancreatic β-cells, overwhelming their folding capacity and leading to a chronic hyper-activation of the UPR. However, further ER stress can lead to β-cell dysfunction and death. DM type 2 is developed in patients that are unable to sustain the UPR of pancreatic β-cells (Fonseca 2009; Ozcan 2012; Bravo 2013).

The pancreatic β-cells have developed an ER compartment that enables efficient secretion of proteins. To promote proper folding and prevent aggregation of misfolded proteins, these cells overexpress several chaperones and foldases, including BiP and PDIA1 (Cadavez 2014). Down-regulation studies have shown that PDIA1 enhances the export of proinsulin, suggesting a putative role as an unfoldase, and thus favoring proinsulin retention within the ER of pancreatic β-cells. However, further studies are needed to analyze its contribution to the limited production of insulin in DM type 2 (Rajpal 2012).

## 6.5. Cardiovascular diseases

Within the vasculature, several thiol isomerases, including PDIA1, have been found on the cell surface of platelets and endothelial cells (Flaumenhaft 2013). Although other thiol isomerases can also be secreted, only three PDI enzymes have proved being implicated in the initiation of thrombus formation *in vivo*: PDIA1, PDIA3, and PDIA6 (Furie 2014). Interestingly, the use of different vascular injury models has shown that fibrin generation dependents on the presence of PDIA1. For instance, after vascular injury, PDIA1 is first secreted from the endothelium and then from the bound platelets. Conversely, in its absence neither platelets aggregate nor fibrin forms in the injured vascular bed. Though is essential for thrombus formation, it is not for maintaining hemostasis. Since thiol isomerase inhibition blocks both platelet accumulation and fibrin generation, specific inhibitors could be helpful for the setting of arterial or venous thrombosis. However, the challenge will be to identify those enzyme-substrate interactions that are critical for thrombus formation (Flaumenhaft 2013; Furie 2014).

## 6. PDIA1/P4HB AS A DRUG TARGET

PDIA1 is ubiquitously expressed in almost all tissues and is highly abundant. It has been estimated that it may account for up to 0.8% of total cellular protein (Galligan 2012). In addition to its physiological function, several studies have shown experimental and clinical evidence suggesting association between defective production and function with development of human pathologies (Uehara 2006; Cho 2008; Hoffstrom 2010; Khan 2011; Chen 2013; Sharda 2015). Therefore, PDIA1 has attracted increasing attention as a potential drug target.

The antibiotic bacitracin was reported to be an inhibitor of PDI in 1981 (Roth 1981). Since then, it has been extensively employed to determine the biochemical and cellular role of PDIA1. Nonetheless, its clinical application is not likely, because it exhibits high nephrotoxicity (Wang 2008) and lacks specificity (Karala 2010). A second antibiotic, ribostamycin, was reported in 2001 as PDIA1 inhibitor that suppresses the chaperone-like activity (Horibe 2001). Interestingly, it binds specifically to another region that is distinct from the active site for isomerase activity (Horibe 2002). Ever since, it has been used to study the chaperone-like activity of PDI enzymes (Ko 2004; Gaucci 2008; Ben Khalaf 2012; Taylor 2014). To date, several PDIA1 inhibitors have been reported (Table 2).

The compound denoted as 16F16 has shown the ability to prevent apoptosis induced by a mutant huntingtin protein, by inhibiting the enzymatic activity (Hoffstrom 2010). Another molecule, named PACMA 31, has exhibited anti-cancer activity in a murine model of human ovarian cancer. This inhibitor also revealed tumor-specific targeting ability (Xu 2012). Additionally, the molecule called P1 has demonstrated ability to inhibit the enzymatic activity both *in vivo* and *in vitro* (Ge 2013). Despite their effectiveness in blocking PDIA1, these inhibitors have not fulfilled the desired pharmacological features, as they have shown low specificity and strong irreversibility (Eirich 2014). Instead, the reversible inhibitor called T8 has served as sensitizing agent in combination with a toxic chemotherapeutic drug to address the vulnerability of various cancer cells and induce apoptosis. Moreover, the target specificity was established by proteomics analysis and further in-depth chemical and biological studies confirmed the unique mode of action through reversible inhibition (Eirich 2014).

**Table 2. Inhibitors used to suppress the activity of human PDIA1/P4HB**

| Inhibitor | Application | $IC_{50}$ (μM) | Origin | Reference |
|---|---|---|---|---|
| Bacitracin | PDI inhibitor | 90 | Natural antibiotic | Roth 1981<br>Xu, 2014 |
| P1 | Inhibition of cancer cells | 1.7 | Small synthetic molecule | Ge, 2013 |
| DTNB | Reaction with thiols. Inhibition of the HIV binding to CD4 receptor | 2600 | Synthetic molecule | Lara 2011 |
| pCMBS | Reaction with thiols | N/A[2] | Small synthetic molecule | Robinson 2006 |
| PAO | Reaction with CXXC motif | 85 | Small synthetic molecule | Bennet 2000 |
| Estrogens | Inhibition of reductase and isomerase activities | 1 | Hormones | Tsibris 1989 |
| RB-11-ca | PDI inhibitor | 23.9 | Small synthetic molecule | Banerjee 2013 |
| 16F16 | Apoptosis inhibitor | 74 | Small synthetic molecule | Hoffstrom 2010<br>Ge, 2013 |
| PACMA 31 | Cell cycle regulator, inhibition of apoptotic cells | 10 | Small synthetic molecule | Xu, 2012 |
| T8 | PDI inhibition in cancer cells | > 300 | Small synthetic molecule | Eirich 2014 |
| Juniferdin | Blocks HIV fusion to CD4 receptor in T-lymphocyte | 0.165 | Natural product | Khan 2011 |
| Quercetin-3-rutinoside | Inhibits thrombus formation | 6.1 | Natural Product | Jasuja 2012 |
| Ribostamicyn[1] | Inhibition of the chaperone activity | N/A[2] | Antibiotic | Horibe 2001 |

[1]At 100:1 ratio, ribostamycin:PDI, complete inhibition of the chaperone activity was detected.

[2]NA, Not Available.

The potential application of PDIA1 as a target for pharmacological control of several human diseases has been well-stablished. However, some issues regarding this protein should be considered, as its ubiquitous distribution and its critical intracellular function. For instance, chronic PDI-silencing is toxic in cultured cells and no PDI-deficient animal model has been developed so far (Jasuja 2012). Furthermore, the precise role of PDIA1 in human diseases remains to be fully clarified (Xu 2014). Hence, new or improved inhibitors with high specificity and low toxicity are required to demonstrate the feasibility of targeting PDIA1 *in vivo*, and thus overcome the significant safety issues before their clinical use (Jasuja 2012).

## Conclusion

Human PDIA1 protein belongs to a large family of 21 members. Its main function is related to the correct oxidative folding of polypeptide substrates, through the proper formation of disulfide bonds with the assistance of its chaperone-like activity. It has further demonstrated its participation in the regulation of homeostasis, as a key component of the unfolded protein response mechanism and the ER-associated degradation process. Recently, the altered function and production of PDIA1 have been related to human pathologies caused by the aggregation of misfolded proteins, e.g., neurodegenerative diseases and coagulation disorders, among others. Excluding some physiological issues, targeting PDIA1 remains as a reasonable approach for pharmacological control of diseases. Additionally, as some security issues remain unsolved for current PDIA1 inhibitors, it is prevalent the need to identify new or improved inhibitors with potential application in clinical trials.

## References

Aebersold R. Constellations in a cellular universe. *Nature*. 2003; 422 (6928): 115-6.

Ali Khan H, Mutus B. Protein disulfide isomerase a multifunctional protein with multiple physiological roles. *Front. Chem.* 2014; 2:70.

Appenzeller-Herzog C, Ellgaard L. The human PDI family: versatility packed into a single fold. *Biochim. Biophys. Acta.* 2008; 1783 (4): 535-48.

Araki K, Iemura S, Kamiya Y, Ron D, Kato K, et al. Ero1-α and PDIs constitute a hierarchical electron transfer network of endoplasmic reticulum oxidoreductases. *J. Cell Biol.* 2013; 202 (6): 861-74.

Banerjee R, Pace NJ, Brown DR, Weerapana, E. 1,3,5-Triazine as a Modular Scaffold for Covalent Inhibitors with Streamlined Target Identification. *J. Am. Chem. Soc.* 2013; 135 (7): 2497-500.

Ben Khalaf N, Muylder G, Louzir H, McKerrow J, Chenik M. Leishmania major protein disulfide isomerase as drug target: enzymatic and functional characterization. *Parasitol. Res.* 2012; 110 (5): 1911-7.

Benham AM. The protein disulfide isomerase family: key players in health and disease. *Antioxid. Redox Signal.* 2012; 16 (8): 781-9.

Bi S, Hong PW, Lee B, Baum LG. Galectin-9 binding to cell surface protein disulfide isomerase regulates the redox environment to enhance T-cell migration and HIV entry. *Proc. Natl. Acad. Sci. U.S.A.* 2011; 108 (26):10650-5.

Bobrovnikova-Marjon E, Grigoriadou C, Pytel D, Zhang F, et al. PERK promotes cancer cell proliferation and tumor growth by limiting oxidative DNA damage. *Oncogene.* 2010; 29 (27): 3881-95.

Bravo R, Parra V, Gatica D, Rodríguez AE, et al. endoplasmic reticulum and the unfolded protein response: dynamics and metabolic integration. *Int. Rev. Cell Mol. Biol.* 2013; 301: 215-90.

Cadavez L, Montane J, Alcarraz-Vizan G, Visa M, et al. Chaperones ameliorate beta cell dysfunction associated with human islet amyloid polypeptide overexpression. *PLOS one.* 2014; 9 (7): e101797.

Cai H, Wang CC, Tsou CL. Chaperone-like activity of protein disulfide isomerase in the refolding of a protein with no disulfide bonds. *J. Biol. Chem.* 1994; 269 (40): 24550-2.

Cerutti N, Killick M, Jugnarain V, Papathanasopoulos M, Capovilla A. Disulfide reduction in CD4 domain 1 or 2 is essential for interaction with HIV glycoprotein 120 (gp120), which impairs thioredoxin-driven CD4 dimerization. *J. Biol. Chem.* 2014; 289 (15): 10455-65.

Chen L, Xu S, Liu L, Wen X, Xu Y, Chen J, Teng J. Cab45S inhibits the ER stress-induced IRE1-JNK pathway and apoptosis via GRP78/BiP. *Cell Death Dis.* 2014; 5: e1219.

Chen X, Zhang X, Li C, Guan T, et al. S-nitrosylated protein disulfide isomerase contributes to mutant SOD1 aggregates in amyotrophic lateral sclerosis. *J. Neurochem.* 2013; 124 (1): 45-58.

Cho J, Furie BC, Coughlin SR, Furie B. A critical role for extracellular protein disulfide isomerase during thrombus formation in mice. *J Clin. Invest.* 2008; 118 (3): 1123-31.

Coe H, Michalak M. ERp57, a multifunctional endoplasmic reticulum resident oxidoreductase. *Int. J. Biochem. Cell Biol.* 2010; 42 (6): 796-9.

Eirich J, Braig S, Schyschka L, Servatius P, et al. A small molecule inhibits protein disulfide isomerase and triggers the chemosensitization of cancer cells. *Angew. Chem. Int. Ed. Engl.* 2014; 54 (47): 12960-5.

Flaumenhaft R. Protein disulfide isomerase as an antithrombotic target. *Trends Cardiovasc. Med.* 2013; 23 (7): 264-8.

Fonseca SG, Burcin M, Gromada J, Urano F. Endoplasmic reticulum stress in beta cells and development of diabetes. *Curr. Opin. Pharmacol.* 2009; 9 (6): 763-70.

Furie B, Flaumenhaft R. Thiol isomerases in thrombus formation. *Circ. Res.* 2014; 114 (7): 1162-73.

Galligan JJ, Petersen DR. The human protein disulfide isomerase gene family. *Hum. Genomics.* 2012; 6 (1): 6.

Gardner BM, Pincus D, Gotthardt K, Gallagher CM, Walter P. Endoplasmic Reticulum Stress Sensing in the Unfolded Protein Response. *Cold Spring Harb. Perspect. Biol. 2013*; 5 (3): a013169.

Gaucci E, Chichiarelli S, Grillo C, Del Vecchio ED, et al. The Binding of Antibiotics to ERp57/GRP58. *J. Antibiot (Tokyo).* 2008; 61 (6): 400-2.

Ge J, Zhang CJ, Li L, Chong LM, et al. Small molecule probe suitable for *in situ* profiling and inhibition of protein disulfide isomerase. *ACS Chem. Biol.* 2013; 8 (11): 2577-85.

Gidalevitz T, Kikis EA, Morimoto RI. A cellular perspective on conformational disease: The role of genetic background and proteostasis networks. *Curr. Opin. Struct. Biol.* 2010; 20 (1): 23–32.

Grek C, Townsend DM. Protein disulfide isomerase superfamily in disease and the regulation of apoptosis. *Endoplasmic Reticulum Stress Dis.* 2014; 1 (1): 4-17.

Halloran M, Parakh S, Atkin JD. The role of s-nitrosylation and s-glutathionylation of protein disulphide isomerase in protein misfolding and neurodegeneration. *Int. J. Cell Biol.* 2013; 2013: 797914.

Hartl FU, Bracher A, Hayer-Hartl M. Molecular chaperones in protein folding and proteostasis. *Nature.* 2011; 475: 324-32.

Hatahet F, Ruddock LW. Protein disulfide isomerase: a critical evaluation of its function in disulfide bond formation. *Antioxid. Redox Signal.* 2009; 11 (11): 2807-50.

Hipp MS, Park SH, Hartl FU. Proteostasis impairment in protein-misfolding and -aggregation diseases. *Trends Cell Biol.* 2014; 24 (9): 506-14.

Hoffstrom BG, Kapplan A, Letso R, Schmid RS, et al. Inhibitors of protein disulfide isomerase suppress apoptosis induced by misfolded proteins. *Nat. Chem. Biol.* 2010; 6 (12): 900-6.

Horibe T, Nagai H, Matsui H, Hagiwara Y, Kikuchi M. Aminoglycoside antibiotics bind to protein disulfide isomerase and inhibit its chaperone activity. *J. Antibiot.* 2002; 55 (5): 528–30.

Horibe T, Nagai H, Sakakibara K, Hagiwara Y, Kikuchi M. Ribostamycin inhibits the Chaperone Activity of Protein Disulfide Isomerase. *Biochem. Biophys. Res. Commun.* 2001; 289 (5): 967-72.

Imaoka S. Chemical stress on protein disulfide isomerases and inhibition of their functions. *Int. Rev. Cell. Mol Biol.* 2011; 290: 121-66.

Jasuja R, Passam FH, Kennedy DR, Kim SH, et al. Protein disulfide isomerase inhibitors constitute a new class of antithrombotic agents. *J. Clin. Invest.* 2012; 122 (6): 2104-13.

Karala AR, Ruddock LW. Bacitracin is not a specific inhibitor of protein disulfide isomerase. *FEBS J.* 2010; 277 (11): 2454-62.

Khan MM, Simizu S, Lai NS, Kawatani M, *et al.* Discovery of a Small Molecule PDI Inhibitor That Inhibits Reduction of HIV-1 Envelope Glycoprotein gp120. *ACS Chem. Biol.* 2011; 6 (3): 245-51.

Kim YE, Hipp MS, Bracher A, Hayer-Hartl M, Hartl FU. Molecular chaperone functions in protein folding and proteostasis. *Annu. Rev. Biochem.* 2013; 82: 323-55.

Klappa P, Ruddock LW, Darby NJ, Freedman RB. The b' domain provides the principal peptide-binding site of protein disulfide isomerase but all domains contribute to binding of misfolded proteins. *EMBO J.* 1998; 17 (4): 927-35.

Ko MK, Kay EP. PDI-mediated ER retention and proteasomal degradation of procollagen I in corneal endothelial cells. *Exp. Cell. Res.* 2004; 295 (1): 25-35.

Kozlov G, Määttänen P, Thomas DY, Gehring K. A structural overview of the PDI family of proteins. *FEBS J.* 2010; 277 (19): 3924-36.

Lara H, Ixtepan-Turrent L, Garza-Treviño EN, Flores-Treviño SM, et al. Antiviral propierties of 5,5'-dithiobis-2-2-nitrobenzoic acid and bacitracin against T-tropic human immunodeficiency virus type 1. *Virol. J.* 2011; 8: 137.

Laurindo FR, Pescatore LA, Fernandes Dde C. Protein disulfide isomerase in redox cell signaling and homeostasis. *Free Radic. Biol. Med.* 2012; 52 (9): 1954-69.

Li H, Korennykh AV, Behrman SL, Walter P. Mammalian endoplasmic reticulum stress sensor IRE1 signals by dynamic clustering. *Proc. Natl. Acad. Sci. U.S.A.* 2010; 107 (37): 16113-8.

Lin JH, Li H, Zhang Y, Ron D, Walter P. Divergent Effects of PERK and IRE1 Signaling on Cell Viability. *PLoS ONE.* 2009; 4 (1): e4170.

Lu J, Holmgren A. The thioredoxin superfamily in oxidative protein folding. *Antioxid. Redox Signal.* 2014; 21 (3): 457-70.

Luo B, Lee AS. The critical roles of endoplasmic reticulum chaperones and unfolded protein response in tumorigenesis and anti-cancer therapies. *Oncogene.* 2013; 32 (7): 805-18.

Mu TW, Ong DS, Wang YJ, Balch WE, *et al.* Chemical and biological approaches synergize to ameliorate protein-folding diseases. *Cell.* 2008; 134 (5): 769-81.

Ozcan L, Tabas I. Role of endoplasmic reticulum stress in metabolic disease and other disorders. *Annu. Rev. Med.* 2012; 63: 317-28.

Pan S, Chen HH, Correia C, Dai H, *et al.* Cell surface protein disulfide isomerase regulates natriuretic peptide generation of cyclic guanosine monophosphate. *PLoS ONE.* 2014; 9 (11): e112986.

Parakh S, Atkin JD. Novel roles for protein disulphide isomerase in disease states: a double edged sword? *Front. Cell Dev. Biol.* 2015; 3: 30.

Pihlajaniemi T, Helaakoski T, Tasanen K, Myllyla R, *et al.* Molecular cloning of the beta-subunit of human prolyl 4-hydroxylase. This subunit and protein disulphide isomerase are products of the same gene. *EMBO J.* 1987; 6 (3): 643-9.

Rajpal G, Schuiki I, Liu M, Volchuk A, Arvan P. Action of protein disulfide isomerase on proinsulin exit from endoplasmic reticulum of pancreatic β-cells. J. Biol. Chem. 2012; 287 (1): 43-7.

Robinson A, O'Neill S, Kiernan A, O'Donoghue N, Moran N. Bacitracin reveals a role for multiple thiol isomerases in platelet function. *Br. J. Haematol.* 2006; 132 (3): 339-48.

Roth RA. Bacitracin: an inhibitor of the insulin degrading activity of glutathione-insulin transhydrogenase. *Biochem. Biophys. Res. Commun.* 1981; 98 (2): 431-8.

Sanderson TH, Gallaway M, Kumar R. Unfolding the unfolded protein response: unique insights into brain ischemia. *Int. J. Mol. Sci.* 2015; 16 (4): 7133-42.

Sharda A, Kim SH, Jasuja R, Gopal S, *et al.* Defective PDI release from platelets and endothelial cells impairs thrombus formation in Hermansky-Pudlak syndrome. *Blood.* 2015; 125 (10): 1633-42.

Shishkin SS, Eremina LS, Kovalev LI, Kovaleva MA. AGR2, ERp57/GRP58, and some other human protein disulfide isomerases. *Biochemistry (Mosc).* 2013; 78 (13): 1415-30.

Stahl S, Burkhart JM, Hinte F, Tirosh B, et al. Cytomegalovirus downregulates IRE1 to repress the unfolded protein response. *PLoS Pathog.* 2013; 9 (8): e1003544.

Stantchev TS, Paciga M, Lankford CR, Schwartzkopff F, *et al.* Cell-type specific requirements for thiol/disulfide exchange during HIV-1 entry and infection. *Retrovirology.* 2012; 9: 97.

Stolf BS, Smyrnias I, Lopes LR, Vendramin A, et al. Protein disulfide isomerase and host-pathogen interaction. *Scientific World Journal.* 2011; 11: 1749-61.

Taylor M, Burress H, Banerjee T, Ray S, *et al.* Substrate-induced unfolding of protein disulfide isomerase displaces the cholera toxin A1 subunit from its holotoxin. *PLOS Pathog.* 2014; 10 (2): e1003925.

Tsibris JC, Hunt LT, Ballejo G, Barker WC, et al. Selective inhibition of protein disulfide isomerase by estrogens. *J. Biol. Chem.* 1989; 264 (24): 13967-70.

Uehara T, Nakamura T, Yao D, Shi ZQ, *et al.* S-Nitrosylated protein-disulphide isomerase links protein misfolding to neurodegeneration. *Nature.* 2006; 441 (7092): 513-7.

Walker AK, Atkin JD. Mechanism of neuroprotection by protein disulphide isomerase in amyotrophic lateral sclerosis. *Neurol. Res. Int.* 2011; 2011: 317340.

Wang C, Chen S, Wang X, Wallis AK, *et al.* Plasticity of human protein disulfide isomerase: evidence for mobility around the x-linker region and its functional significance. *J. Biol. Chem.* 2010; 285 (35): 26788-97.

Wang C, Li W, Ren J, Fang J, et al. Structural insights into the redox-regulated dynamic conformations of human protein disulfide isomerase. *Antioxid. Redox Signal.* 2013; 19 (1): 36-45.

Wang C, Yu J, Huo L, Wang L, *et al.* Human protein-disulfide isomerase is a redox-regulated chaperone activated by oxidation of domain a'. *J. Biol. Chem.* 2012; 287 (2): 1139-49.

Wang EJ, Snyder RD, Fielden MR, Smith RJ, Gu YZ. Validation of putative genomic biomarkers of nephrotoxicity in rats. *Toxicology.* 2008; 246 (2-3): 91-100.

Xu S, Butkevich AN, Yamada R, Zhou Y, et al. Discovery of an orally active small-molecule irreversible inhibitor of protein disulfide isomerase for ovarian cancer treatment. *Proc. Natl. Acad. Sci. U.S.A.* 2012; 109 (40): 16348-53.

Xu S, Sankar S, Neamati N. Protein disulfide isomerase: a promising target for cancer therapy. *Drug Discov. Today.* 2014; 19 (3): 222-40.

Zhou J, Wu Y, Wang L, Rauova L, et al. The disulfide isomerase ERp57 is required for fibrin deposition *in vivo*. *J. Thromb. Haemost.* 2014; 12: 1890–7.

In: New Research on Molecular Chaperones
Editor: Jaime Wyatt
ISBN: 978-1-63483-543-5


*Chapter 3*

# Cytoplasmic and Nuclear Functions for the Prefoldin-like URI/Bud27

***Verónica Martínez-Fernández, Ana Isabel Garrido-Godino, Abel Cuevas-Bermúdez and Francisco Navarro****

Departamento de Biología Experimental, Facultad de Ciencias Experimentales, Universidad de Jaén, Paraje de las Lagunillas, Jaén, Spain

## Abstract

The unconventional prefoldin Rpb5 interactor (URI/RMP), and its ortholog in yeast, Bud27, are members of the prefoldin (PFD) family of ATP-independent molecular chaperones. URI was originally identified as a protein binding the Rpb5 subunit of all three nuclear RNA polymerases (RNA pol) and is considered to function as a scaffold protein able to assemble additional members of the PFD family in both human and yeast.

URI and Bud27 have been linked to essential cellular processes both in the nucleus and the cytoplasm, such as RNA polymerases assembly, translation initiation, genome stability, DNA damage response, and viability. Furthermore, URI has been shown to be involved in several types of cancer processes and to regulate the TOR pathway, not only in humans, but also in yeast.

* Corresponding author: Francisco Navarro. E-mail: fngomez@ujaen.es. Tel: 00-34-953-212771. Fax: 00-34-953-211875.

Taken together, this data points to a role for URI/Bud27 as a global molecular chaperone participating in multiple essential cellular processes.

In this review, we focus on the cytoplasmic and nuclear functions of URI/Bud27 in different organisms.

## STRUCTURE AND LOCALIZATION OF URI/BUD27

The URI gene was first isolated and cloned from a human HepG2 cDNA library [1], although other human variants were later described [2, 3]. The corresponding gene from *Drosophila melanogaster* is essential for viability, while mutation of *Caenorhabditis elegans* URI, *uri*-1, leads to sterility as well as to multiple and variable somatic defects [4, 5]. The yeast ortholog *BUD27* is not essential for growth but the budding yeast deletion strain is defective for the expression of genes in the nutrient response pathway [6].

Human URI and *S. cerevisiae* ortholog Bud27 reveal three highly conserved regions, termed the PFD domain, the RPB5-binding region and the URI box (See Figure 1) [2, 6, 7]. In addition, an Asp-rich region, including 13 contiguous Asp residues, is present in the central region of URI, with a high degree of homology to several unrelated proteins with different functions [1]. URI conserved PFD domain, mediates the binding of URI with Pfd6 [6], a component of the PFD/GimC complex that functions as a chaperone to assist in the folding of newly synthesized actin and tubulin [8, 9].

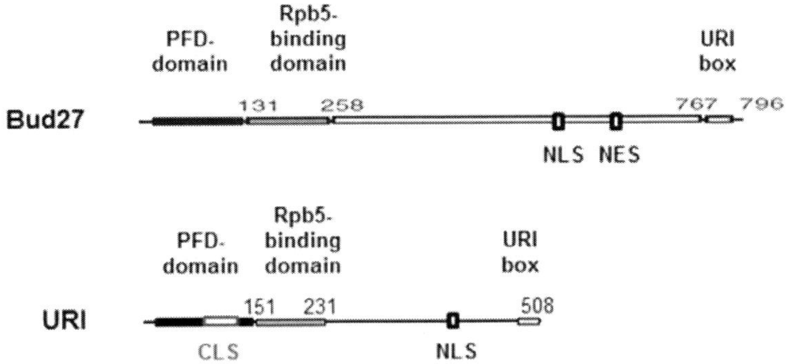


Figure 1. Structure of human URI and *S. cerevisiae* ortholog Bud27 revealing three conserved regions, such as the PFD domain, the RPB5-binding region and the URI-box. NLS: nuclear localization signal. NES: Nuclear export signal. CLS: cytoplasmic localization signal.

Localization of the unconventional prefoldin Rpb5 interactor (URI/RMP), and its orthologs has been extensively study. In human cells, URI shuttles between cytoplasm and nucleus, although it is localized mainly in the cytoplasm [2]. This is consistent with the fact that in mammalian cells URI contains two localization signals, a nuclear localization signal (NLS), aa 339 to 344, and a cytoplasmic localization signal (CLS) in its amino-terminal region, aa 88-118 [2] (Dorjsuren 1998 [1]). In addition, it also agrees with the cytoplasmic localization of the N-terminally deleted NNX3 clone of URI/RMP [3]. Delgermaa et al. isolated a DNA methyltransferase 1 (DNMT1)-associating protein (DMAP1), that promotes the nuclear localization of URI by interacting with the CC domain (Coiled-coil domain that functions as CLS) [2] and a mass spectrometry analysis of nuclear extracts in human prostate cells, identify URI as a part of the R2TP/prefoldin-like complex [10]. Furthermore, the mechanisms governing the URI cytoplasmic/nuclear translocation in human occur through the CRM1/XPO1-1 exportin complex previously shown also to mediate nuclear export of RNA polymerase subunits in humans [10]. *Drosophila* Uri, which is a predominantly cytoplasmic protein with perinuclear localization, contains four NLS sequences [5]. Furthermore, the existence of two NLS sequences in worm uri-1 suggests that it is also a shuttling protein [5].

Although the yeast ortholog Bud27 was first described as a cytoplasmic protein [11], the deletion of its NES sequence led to a nuclear accumulation [7]. The sequence analysis of Bud27 predicted a possible leucine-rich nuclear export signal, NES, between positions 686 and 695 (LRDEIRDFQL) and a nuclear localization signal, NLS, (amino acids 562 to 595) [7]. Notably, in contrast to human URI, Bud27 shuttling is independent of Xpo1 exportin and of the microtubule polarization [7, 10].

Finally, in human cells, URI can be phosphorylated or acetylated at different sites. Although the impact of these posttranslational modifications is not well understood, they could be important for URI functions [10].

## CYTOPLASMIC ROLES OF URI/BUD27

The main cytoplasmic localization of URI/Bud27 in different organisms accounts for a participation of these proteins in cytoplasmic processes.

## URI Could Participate in RNA Polymerase II Assembly

A putative role for human URI in RNA pol II assembly has been suggested [7, 12, 13]. URI, identified as a protein binding the RPB5 common subunit to all three nuclear RNA polymerases in both human and yeast [1, 6, 7] is considered to function as a scaffold protein able to assemble additional members of the PFD family in human as well as in yeast. Human and yeast PFD is a complex composed of six different subunits (PFD1-PFD6) referred to as the prefoldin/GimC complex, which functions as a molecular chaperone and delivers newly synthesized unfolded proteins to cytosolic chaperonin TRiC/CCT to facilitate their folding [12-14]. In humans, PFDs and URI are constituents of an 11-subunit complex associated with RNA pol II, namely the RPAP3/R2TP/prefoldin-like complex, suggested to participate in RNA pol II assembly [15, 16]. This prefoldin-like complex comprises the RNA pol II subunit POLR2E (RPB5) and the human orthologs of the yeast R2TP complex, including RPAP3, PIH1D1, RUVBL1, and RUVBL2. In addition, WDR92, which interacts with RPAP3, is found to contain the two subunits shared with the canonical prefoldin complex PFDN2 and PFDN6 (HKE2), and three prefoldin-like proteins, UXT (ART-27), PDRG1, and C19orf2 (URI, RMP) [10, 12, 17]. Notably, the yeast R2TP protein Tah1 (RPAP3) has been characterized as HSP90 co-chaperone [18]. Furthermore, the human hSpagh (RPAP3), which is an HSP90 cofactor [19], mediates the assembly of RNA pol II in the cytoplasm [15] in coordination with HSP90. Finally, other proteins suggested to be transient interactors, but not components of the stable complex, co-purified with some tagged subunits of the RPAP3 (R2TP/Prefoldin-like) complex, such as POLR2A (RPB1), RPAP2, RPAP4, and RP11-529I10.4 (DPCD) [17].

## Bud27 Mediates the Biogenesis of the Three Nuclear RNA Polymerases in Yeast

Bud27 is the first protein demonstrated to participate in the biogenesis of the three RNA polymerases in *S. cerevisiae*, modulating their assembly by stabilizing the interaction between Rpb5 and the rest of the RNA pols complexes [7]. Data from proteomic analysis in yeast showed that Bud27 interacts with subunits of the three RNA polymerases, including Rpa190, Rpa135, and Rpa49 (RNA pol I); Rpc160 and Rpc128 (pol III); Rpc40 (pol I and III); Rpb1 (pol II); Rpb10 and Rpb5 (pol I, II and III) [7]. These data, in agreement with a putative role of human URI in RNA pol II assembly [12], also suggest a role for Bud27 in RNA pols biogenesis. In accordance with this, the deletion of *BUD27* results in a cytoplasmic accumulation of the three RNA polymerases, but their nuclear localization seems not to depend on Bud27 localization, excluding a role of this prefoldin-like complex in RNA polymerases transport [7]. The role of Bud27 in RNA pols biogenesis depends on Rpb5 and is necessary for the correct association of Rpb5 and Rpb6 to RNA pol I, II, and III [7]. This process occurs in a latter step of the biogenesis before their translocation to the nucleus, and accounts for a role of Bud27 as a chaperone, in agreement with its role in protein folding [13]. In fact in *bud27Δ* cells, the stoichiometry of Rpb5 and Rpb6 to other subunits of RNA pol I, II, and III decreases, a process that is overcome by increasing the dosage of Rpb5. Thus, the deletion of the Rpb5-binding domain of Bud27 leads to a cytoplasmic accumulation of the three RNA pols, this domain not being required for growth [7]. Notably, the PFD domain is necessary neither for nuclear RNA pol localization, nor for growth. In addition, the defect in Rpb6 assembly is consistent with the physical contact between Rpb5 and Rpb6 on the RNA pol II structure [20].

The role of Bud27 in RNA pols biogenesis seems to be independent of the rest of the prefoldin complex. YKE2 (PFD6), which physically interacts with Bud27 [7, 13], with Rpb5 [21], and genetically with other elements of RNA pol I and III machinery [22, 23], GIM6 (PFD1), which genetically interacts with BUD27 [11] and with subunits of RNA pol I and III [22, 24, 25], or GIM4 (PFD2) are dispensable for the nuclear localization of the three RNA pols [7]. In addition, the function of Bud27 in RNA pols biogenesis is independent of Iwr1 [7], a factor that binds Rpb1 and Rpb2 and directs its nuclear import [26, 27], despite that Iwr1 genetically interacts with *BUD27* and *RPB5* [28, 29].

Finally, the silencing of URI with siRNA in human pulmonary fibroblast resulted in the accumulation of RPB1 in the cytoplasm of treated cells, suggesting that, as is the case for Bud27, URI has a role in the biogenesis of at least RNA pol II [7].

## Bud27 Participates in Translation

Protein synthesis is a highly regulated process that coordinates the chain synthesis carried out by ribosomes and translation factors with cotranslational folding of nascent polypeptides by chaperone complexes (reviewed in [30]).

Nascent proteins are targets of post- and co-translational quality-control mechanisms, including the correct folding of the chains assisted by a large set of prefoldins and chaperons, as well as a mechanism which involves the Gcn2-dependent phosphorylation of the eIF2a translation factor. This phosphorylation inhibits translation of most proteins and is required to activate the translation of a few stress-regulated proteins, such as the transcription factors ATF4 in mammals and Gcn4 in yeast (reviewed in [31, 32]).

Bud27 is required to repress Gcn4 translation under nutrient-rich conditions since *bud27Δ* shows a $Gcd^-$ (general control derepressed) mutant phenotype [11]. The $Gcd^-$ mutants were identified to decrease the amount of ternary complex, TC (consisting of the initiator methionyl-tRNA, Met-$tRNA_i^{Met}$, the initiation factor eIF2 and GTP), bound to the 40S subunit, and therefore to increase the Gcn4 translation (reviewed in [33]). In accordance with the fact that defects in protein translation reduce fitness as slow growth and temperature sensitivity, *bud27Δ* cells show slow growth and temperature sensitive phenotypes [34]. In addition, *bud27Δ* cells are hypersensitive to translation inhibitors as cycloheximide, paramomycin and hygromycin B and show an altered polysome profile with an 80S peak higher than in a wild-type strain [11].

The role of Bud27 in translation occurs at the level of translation initiation by promoting efficient recruitment of the TC to 40S subunits, in agreement with the physical interaction between Bud27 and the translation initiation factor eIF1A (encoded by TIF11), which facilitates TC recruitment [11, 13], and with the association between Bud27 and ribosomes [11]. Those data suggest that Bud27 may support translation initiation by directly promoting TC recruitment and/or activating initiation factors such as eIF1A. In addition to eIF1A, other Bud27 interactors, most of them linked to protein synthesis, have been described [13]. Several chaperones interacting with Bud27, such as Ssb1 and Sis1 [13], members of Hsp70 and Hsp40 family, or the GimC-complex component Pfd6 that have been classified as chaperones linked to protein synthesis (CLIPs) [6, 13, 35-38], suggest that Bud27 may not directly participate in protein folding, but rather functions as a regulator of translation and protein quality control [11]. Finally, Bud27 also interacts with the peptidyl-prolyl cis-trans isomerases Fpr1p and Crp1p, which might be involved in the folding of newly synthesized proteins [13].

# NUCLEAR ROLES OF URI/BUD27

URI/ Bud27 has a weak nuclear localization in some organisms such as mammals, *C. elegans*, *Drosophila*, and *S. cerevisiae.* Within the nucleus, URI/ Bud27 physically binds to Rpb5, other nuclear proteins involved in transcription [1, 39], tumor suppression proteins [40], and chromatin remodeling complex RSC [41].

## Nuclear Functions of URI

A role for URI in transcription activation has been proposed. In fact, URI/RMP in humans was reported to regulate transcription and to act as a co-repressor of the activated transcription by competing with HBx (Hepatitis B virus X protein) to bind the general transcription factor IIB (TFIIB) [1]. URI/RMP covers the regions necessary for TFIIB and HBx binding to Rpb5, suggesting that it may interfere with the trimeric interaction of Rpb5, TFIIB, and HBx to counteract HBx transactivation essential for Hepatitis B virus infection.

In addition, the general transcription factor IIF (TFIIF), which assembles in the preinitiation complex and functions in both transcription initiation and elongation, associates with URI *in vivo*. In fact, experiments with COS1 cells suggest that TFIIF is necessary for the URI suppression function of activated transcription [39].

URI controls androgen receptor (AR) transcription, acting as an AR repressor, and is able to modulate AR recruitment to target genes in human cells [42]. Whereas URI overexpression results in decreased AR-transcription, URI depletion increases the expression of androgen-regulated genes. In addition, URI knockdown increases AR recruitment at NKX3.1, an AR-regulated gene, suggesting a direct role of URI in AR transcription regulation. Furthermore, URI was co-immunoprecipitated in nuclear fractions of prostate cells with the androgen receptor-trapped clone 27 (Art-27), an AR co-repressor. In addition, the depletion of Art-27 (also known as UXT) correlated with loss of URI and vice versa [42]. Notably, mass spectrometry analysis of nuclear URI interactors in prostate cells identified URI and Art-27 as part of the R2TP/prefoldin-like complex [10, 15].

Notably, the analysis of URI-associated proteins from unfractionated HEK-293 cell extracts identified the interaction between URI and the human homolog of *S. cerevisiae* Paf1, a complex involved in cell-cycle control, RNA pol II phosphorylation, and histone modification during transcription elongation [40].

However, the role of URI in the nucleus is not limited to transcription, but is also suggested to function in the maintenance of DNA integrity. URI-1 in *C. elegans* is essential to maintain DNA stability in the absence of exogenous DNA damage. In agreement with this contention, *uri*-1 mutants are sterile due to germline proliferation failures caused by the loss of DNA integrity [4]. Furthermore, *Drosophila* $uri^{110b}$ mutants showed DNA damage in testis cells [5]. Accordingly, overexpression of URI in mouse hepatocellular carcinoma is able to provoke tumorigenesis through DNA damage inhibition of *de Novo* NAD + synthesis [43].

## Bud27 Modulates mRNA Biogenesis and RNA Pol II Transcription Elongation in Yeast

Bud27 contributes to mRNA biogenesis and RNA pol II transcription elongation in *S. cerevisiae* [41]. The deletion of *BUD27* exerts an impact on RNA pol II occupancy throughout the genes, probably as a consequence of the lower amount of enzyme transported to the nucleus [7], and of mRNA accumulation. Bud27 associates with the chromatin and with the RNA pol II phosphorylated forms (CTD-Ser5P and CTD-Ser2P). This RNA pol II CTD domain, which is conserved from yeast to human, is regulated by phosphorylation to modulate transcription [44]. In this sense, *bud27Δ* mutant affects the phosphorylated status of RNA pol II [41].

Deletion of *BUD27* affects transcription elongation and *bud27Δ* mutant cells are sensitive to drugs affecting the intracellular nucleotide pools. Furthermore, Bud27 associates *in vivo* with the Sth1 component of chromatin remodeling complex RSC, in accordance with the genetic interactions described between *BUD27* and *RSC1*, *RSC8*, *RSC4* [22]. In addition, Bud27 mediates the association of RSC with RNA pol II. In fact, co-immunoprecipitation analysis revealed that the deletion of *BUD27* affects the association between Sth1 and RNA pol II [41].

Data from Mirón-García et al. [41] suggest that Bud27 not only contributes to the folding of Rpb5 within RNA pol II, thereby modulating its activity, but also could participate in the proper assembly of other chromatin-associated factors, such as the RSC complex.

Finally, Bud27 has been suggested to be involved in DNA-repair processes [45]. Meanwhile, *bud27Δ* cells show high levels of Rad52, a protein involved in the repair of double-strand breaks in the DNA during vegetative growth and meiosis, as well as in inter-homolog recombination. Also, the *bud27Δ* mutant displays an impaired sister chromatid recombination [45].

## URI/BUD27, A COMPONENT OF THE TOR PATHWAY

Cells are sensitive to external signals, such as cytokines, growth factors or different kinds of stress, which result in a cell response by a signaling pathway and later in a specific gene expression. Starvation or lack of nutrients leads to the activation of the mammalian target of rapamycin (mTOR), which is conserved from yeast to human. The TOR signal-transduction pathway is a key mechanism that occupies a cell growth and proliferative response. In addition, mTOR signaling controls cell growth regulating transcription, translation, protein kinase C signaling, protein degradation, membrane traffic, and actin regulation [46].

URI and Bud27 reportedly participate in the regulation of nutrient-sensitive, TOR-dependent transcription programs [6]. The genome-wide expression analysis in a *bud27Δ* mutant showed 39 genes whose expression was affected by the loss of Bud27, all of them involved in amino acid metabolism [6]. Notably, 29 out of the 39 ORFs affected in the *Δbud27* mutant overlapped with the 331 ORFs altered in both rapamycin and amino acid starvation conditions in yeast cells. A large fraction of genes altered under Bud27 deficiency contains a consensus-binding site for Gcn4, a transcription factor that activates the amino acid biosynthesis genes under starvation conditions [47]. However, not all *BUD27*-dependent genes are *GCN4*-dependent, such as some tRNA genes and *GAP1* (rapamicyn-induced gene). Furthermore, the fact that the levels of Gcn4 are increased in *bud27Δ* cells suggests a role for Bud27 in the TOR pathway through Gcn4 regulation.

Previous studies have shown that in human cells URI is phosphorylated in response to stimuli downstream of the mTOR pathway [6, 48]. URI is phosphorylated by the p70S6 kinase in an mTOR-dependent manner, in human prostate cells, a process modulating the inhibition/activation of the mTOR pathway, integrating extracellular and metabolic stimuli with transcriptional regulation by the androgen receptor [42].

The TOR pathway is involved in growth-related processes, including ribosome biogenesis [46]. Ribosome synthesis requires the coordinated activities of RNA polymerase I, for the synthesis of some rRNAs, RNA polymerase II for ribosomal proteins genes and Pol III for the synthesis of 5S rRNA. The physical interaction of URI/Bud27 with Rpb5, a shared subunit of all three nuclear polymerases, suggests a conserved role for URI/Bud27 as a target of the nutrient-signaling pathway in mTOR-dependent control of gene expression.

## URI AND CANCER

Multiple studies have recently described the oncogenic property of URI during the progression of various cancers or cancer-cell lines, i.e., during the progression of hepatocellular carcinoma, ovarian cancer, and possibly prostate cancer. Most recently, URI has been shown to be up-regulated in cervical cancer, another reproductive-system tumor endometrioid adenocarcinoma [49].

URI is amplified and overexpressed in ovarian cancer-cell lines and human ovarian carcinomas, correlating with aggressive behavior [50]. Furthermore, many studies have reported copy-number gain or amplification of 19q12 locus, containing URI, in a variety of carcinomas [51-57]. Notably, CCNE1, closely related to URI, and found at the 19q12 locus, is also amplified and may contribute to the oncogenic effect [58].

URI in coordination with another set of proteins (S6K1, PP1g, and BAD) constitutes a functional mitochondrial-signaling network, a homeostatic mechanism modulating the cell's mitochondrial threshold, depending on the availability of nutrients and growth factors [48]. This network, regulated by a negative-feedback system through mTOR-S6K1 and by PP1 phosphorylation mediates that URI, affects cell survival [48]. Furthermore, URI coexists with PP1g in a stable mitochondrial complex and forms an integral part of a homeostatic signaling mechanism involved in setting the mitochondrial threshold for apoptosis in accordance with growth-factor and nutritional signals conveyed via the mTOR-S6K1 signaling axis [48].

Overexpressing URI in tumorigenic cell lines enhances resistance of cells to apoptosis in response to death cues [59-61], allowing enhanced cell proliferation and tumor growth *in vivo* [50]. In addition, in some situations, overexpression of URI can cause tumorigenesis through DNA damage inhibition of *de Novo* NAD+ synthesis, as happens in hepatocellular carcinoma (HCC) [43]. In agreement with this, restoring NAD+ pools with nicotinamide riboside (NR) prevents DNA damage and tumor formation [43]. Finally, oncogenic properties of URI by overexpression have also been described in multiple myeloma cells, through the activation of the IL-6/STAT3 pathway.

Notably IL-6 has a particularly important role in myeloma cell proliferation and chemoresistance [62].

## Conclusion

The multiple functions involving URI/Bud27 account for this prefoldin acting as c-chaperone assisting multiple cellular processes both in the nucleus and the cytoplasm (Figure 2).

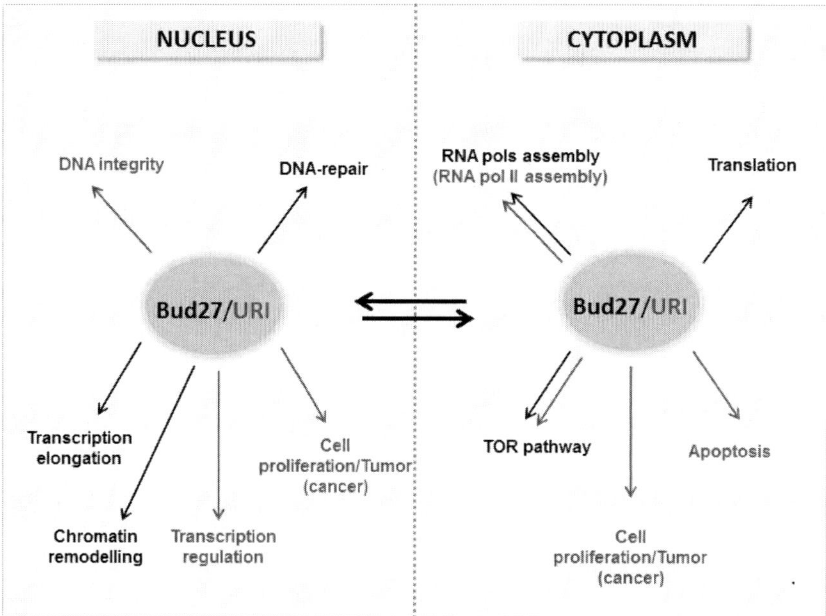


Figure 2. Schematic representation of URI/Bud27 functions in both the nucleus and the cytoplasm. In red those functions described for URI and in black, those demonstrated for Bud27 from *S. cerevisiae.*

## Acknowledgments

This work was supported by grants of Spanish Ministry of Economy and Competitiveness, MINECO, and FEDER (BFU2013-48643-C3-2-P to F.N.), Junta de Andalucía (BIO258, PI10-CVI6521 to F.N.). V. M-F is recipient of

Junta de Andalucía fellowship. A. I. G-G was recipient of a postdoctoral fellowship associated to BFU2013-48643-C3-2-P project. A. C-B is recipient of MINECO fellowship.

## REFERENCES

[1] Dorjsuren D., Lin Y., Wei W., Yamashita T., Nomura T., Hayashi N. et al., RMP, a novel RNA polymerase II subunit 5-interacting protein, counteracts transactivation by hepatitis B virus X protein. *Mol Cell Biol.* 1998 Dec;18(12):7546-55.

[2] Delgermaa L., Hayashi N., Dorjsuren D., Nomura T., Thuy le T. T., Murakami S. Subcellular localization of RPB5-mediating protein and its putative functional partner. *Mol Cell Biol.* 2004 Oct;24(19):8556-66.

[3] Van Leuven F., Torrekens S., Moechars D., Hilliker C., Buellens M., Bollen M. et al., Molecular cloning of a gene on chromosome 19q12 coding for a novel intracellular protein: analysis of expression in human and mouse tissues and in human tumor cells, particularly Reed-Sternberg cells in Hodgkin disease. *Genomics.* 1998 Dec 15;54(3):511-20.

[4] Parusel C. T., Kritikou E. A., Hengartner M. O., Krek W., Gotta M. URI-1 is required for DNA stability in C. elegans. *Development.* 2006 Feb;133(4):621-9.

[5] Kirchner J., Vissi E., Gross S., Szoor B., Rudenko A., Alphey L. et al., Drosophila Uri, a PP1alpha binding protein, is essential for viability, maintenance of DNA integrity and normal transcriptional activity. *BMC Mol Biol.* 2008;9:36.

[6] Gstaiger M., Luke B., Hess D., Oakeley E. J., Wirbelauer C., Blondel M. et al., Control of nutrient-sensitive transcription programs by the unconventional prefoldin URI. *Science.* 2003 Nov 14;302(5648):1208-12.

[7] Miron-Garcia M. C., Garrido-Godino A. I., Garcia-Molinero V., Hernandez-Torres F., Rodriguez-Navarro S., Navarro F. The prefoldin bud27 mediates the assembly of the eukaryotic RNA polymerases in an rpb5-dependent manner. *PLoS Genet.* 2013;9(2):e1003297.

[8] Geissler S., Siegers K., Schiebel E. A novel protein complex promoting formation of functional alpha- and gamma-tubulin. *EMBO J.* 1998 Feb 16;17(4):952-66.

[9] Siegers K., Waldmann T., Leroux M. R., Grein K., Shevchenko A., Schiebel E, et al., Compartmentation of protein folding *in vivo*:

Sequestration of non-native polypeptide by the chaperonin-GimC system. *EMBO J.* 1999 Jan 4;18(1):75-84.

[10] Mita P., Savas J. N., Ha S., Djouder N., Yates J. R., 3rd, Logan S. K. Analysis of URI nuclear interaction with RPB5 and components of the R2TP/prefoldin-like complex. *PLoS One*. 2013;8(5):e63879.

[11] Deplazes A., Mockli N., Luke B., Auerbach D., Peter M. Yeast Uri1p promotes translation initiation and may provide a link to cotranslational quality control. *EMBO J.* 2009 May 20;28(10):1429-41.

[12] Cloutier P., Coulombe B. New insights into the biogenesis of nuclear RNA polymerases? *Biochem Cell Biol.* 2010 Apr;88(2):211-21.

[13] Mockli N., Deplazes A., Hassa P. O., Zhang Z., Peter M., Hottiger M. O. et al., Yeast split-ubiquitin-based cytosolic screening system to detect interactions between transcriptionally active proteins. *Biotechniques*. 2007 Jun; 42(6):725-30.

[14] Millan-Zambrano G., Chavez S. Nuclear functions of prefoldin. *Open Biol.* 2014 Jul;4(7).

[15] Boulon S., Pradet-Balade B., Verheggen C., Molle D., Boireau S., Georgieva M. et al., HSP90 and its R2TP/Prefoldin-like cochaperone are involved in the cytoplasmic assembly of RNA polymerase II. *Mol Cell.* 2010 Sep 24;39(6):912-24.

[16] Forget D., Lacombe A. A., Cloutier P., Al-Khoury R., Bouchard A., Lavallee-Adam M. et al., The protein interaction network of the human transcription machinery reveals a role for the conserved GTPase RPAP4/GPN1 and microtubule assembly in nuclear import and biogenesis of RNA polymerase II. *Mol Cell Proteomics*. 2010 Dec;9 (12):2827-39.

[17] Cloutier P., Al-Khoury R., Lavallee-Adam M., Faubert D., Jiang H., Poitras C. et al., High-resolution mapping of the protein interaction network for the human transcription machinery and affinity purification of RNA polymerase II-associated complexes. *Methods*. 2009 Aug;48 (4):381-6.

[18] Zhao R., Davey M., Hsu Y. C., Kaplanek P., Tong A., Parsons A. B. et al., Navigating the chaperone network: an integrative map of physical and genetic interactions mediated by the hsp90 chaperone. *Cell.* 2005 Mar 11;120(5):715-27.

[19] Boulon S., Marmier-Gourrier N., Pradet-Balade B., Wurth L., Verheggen C., Jady B. E. et al., The Hsp90 chaperone controls the biogenesis of L7Ae RNPs through conserved machinery. *J Cell Biol.* 2008 Feb 11;180(3):579-95.

[20] Cramer P., Bushnell D. A., Kornberg R. D. Structural basis of transcription: RNA polymerase II at 2.8 angstrom resolution. *Science*. 2001 Jun 8;292(5523):1863-76.
[21] Krogan N. J., Cagney G., Yu H., Zhong G., Guo X., Ignatchenko A. et al., Global landscape of protein complexes in the yeast Saccharomyces cerevisiae. *Nature*. 2006 Mar 30;440(7084):637-43.
[22] Collins S. R., Miller K. M., Maas N. L., Roguev A., Fillingham J., Chu C. S. et al., Functional dissection of protein complexes involved in yeast chromosome biology using a genetic interaction map. *Nature*. 2007 Apr 12;446(7137):806-10.
[23] Wilmes G. M., Bergkessel M., Bandyopadhyay S., Shales M., Braberg H., Cagney G. et al., A genetic interaction map of RNA-processing factors reveals links between Sem1/Dss1-containing complexes and mRNA export and splicing. *Mol Cell*. 2008 Dec 5;32(5):735-46.
[24] Ye P., Peyser B. D., Pan X., Boeke J. D., Spencer F. A., Bader J. S. Gene function prediction from congruent synthetic lethal interactions in yeast. *Mol Syst Biol*. 2005;1:2005 0026.
[25] Zheng J., Benschop J. J., Shales M., Kemmeren P., Greenblatt J., Cagney G. et al., Epistatic relationships reveal the functional organization of yeast transcription factors. *Mol Syst Biol*. 2010 Oct 5;6: 420.
[26] Czeko E., Seizl M., Augsberger C., Mielke T., Cramer P. Iwr1 directs RNA polymerase II nuclear import. *Mol Cell*. 2011 Apr 22;42(2):261-6.
[27] Wild T., Cramer P. Biogenesis of multisubunit RNA polymerases. *Trends Biochem Sci*. 2012 Mar;37(3):99-105.
[28] Costanzo M., Baryshnikova A., Bellay J., Kim Y., Spear E. D., Sevier C. S. et al., The genetic landscape of a cell. *Science*. 2010 Jan 22;327 (5964):425-31.
[29] Esberg A., Moqtaderi Z., Fan X., Lu J., Struhl K., Bystrom A. Iwr1 protein is important for preinitiation complex formation by all three nuclear RNA polymerases in Saccharomyces cerevisiae. *PLoS One*. 2011;6(6):e20829.
[30] Frydman J. Folding of newly translated proteins *in vivo*: The role of molecular chaperones. *Annu Rev Biochem*. 2001; 70:603-47.
[31] Dever T. E. Gene-specific regulation by general translation factors. *Cell*. 2002 Feb 22;108(4):545-56.
[32] Gebauer F., Hentze M. W. Molecular mechanisms of translational control. *Nat Rev Mol Cell Biol*. 2004 Oct; 5(10):827-35.

[33] Hinnebusch A. G. Translational regulation of GCN4 and the general amino acid control of yeast. *Annu Rev Microbiol*. 2005;59:407-50.

[34] Cuesta R., Hinnebusch A. G., Tamame M. Identification of GCD14 and GCD15, novel genes required for translational repression of GCN4 mRNA in Saccharomyces cerevisiae. *Genetics*. 1998 Mar;148(3):1007-20.

[35] Albanese V., Yam A. Y., Baughman J., Parnot C., Frydman J. Systems analyses reveal two chaperone networks with distinct functions in eukaryotic cells. *Cell*. 2006 Jan 13;124(1):75-88.

[36] Nelson R. J., Ziegelhoffer T., Nicolet C., Werner-Washburne M., Craig E. A. The translation machinery and 70 kd heat shock protein cooperate in protein synthesis. *Cell*. 1992 Oct 2;71(1):97-105.

[37] Rakwalska M., Rospert S. The ribosome-bound chaperones RAC and Ssb1/2p are required for accurate translation in Saccharomyces cerevisiae. *Mol Cell Biol*. 2004 Oct; 24(20):9186-97.

[38] Zhong T., Arndt K. T. The yeast SIS1 protein, a DnaJ homolog, is required for the initiation of translation. *Cell*. 1993 Jun 18;73(6):1175-86.

[39] Wei W., Gu J. X., Zhu C. Q., Sun F. Y., Dorjsuren D., Lin Y. et al., Interaction with general transcription factor IIF (TFIIF) is required for the suppression of activated transcription by RPB5-mediating protein (RMP). *Cell Res*. 2003 Apr;13(2):111-20.

[40] Yart A., Gstaiger M., Wirbelauer C., Pecnik M., Anastasiou D., Hess D., et al., The HRPT2 tumor suppressor gene product parafibromin associates with human PAF1 and RNA polymerase II. *Mol Cell Biol*. 2005 Jun;25(12):5052-60.

[41] Miron-Garcia M. C., Garrido-Godino A. I., Martinez-Fernandez V., Fernandez-Pevida A., Cuevas-Bermudez A., Martin-Exposito M. et al., The yeast prefoldin-like URI-orthologue Bud27 associates with the RSC nucleosome remodeler and modulates transcription. *Nucleic Acids Res*. 2014 Sep;42(15):9666-76.

[42] Mita P., Savas J. N., Djouder N., Yates J. R., 3rd, Ha S., Ruoff R. et al., Regulation of androgen receptor-mediated transcription by RPB5 binding protein URI/RMP. *Mol Cell Biol*. 2011 Sep;31(17):3639-52.

[43] Tummala K. S., Gomes A. L., Yilmaz M., Grana O., Bakiri L., Ruppen I. et al., Inhibition of *de novo* NAD(+) synthesis by oncogenic URI causes liver tumorigenesis through DNA damage. *Cancer Cell*. 2014 Dec 8;26(6):826-39.

[44] Garcia A., Rosonina E., Manley J. L., Calvo O. Sub1 globally regulates RNA polymerase II C-terminal domain phosphorylation. *Mol Cell Biol.* 2010 Nov;30(21):5180-93.

[45] Munoz-Galvan S., Jimeno S., Rothstein R., Aguilera A. Histone H3K56 acetylation, Rad52, and non-DNA repair factors control double-strand break repair choice with the sister chromatid. *PLoS Genet.* 2013;9(1): e1003237.

[46] Mayer C., Grummt I. Ribosome biogenesis and cell growth: mTOR coordinates transcription by all three classes of nuclear RNA polymerases. *Oncogene.* 2006 Oct 16;25(48):6384-91.

[47] Natarajan K., Meyer M. R., Jackson B. M., Slade D., Roberts C., Hinnebusch A. G. et al., Transcriptional profiling shows that Gcn4p is a master regulator of gene expression during amino acid starvation in yeast. *Mol Cell Biol.* 2001 Jul; 21(13):4347-68.

[48] Djouder N., Metzler S. C., Schmidt A., Wirbelauer C., Gstaiger M., Aebersold R. et al., S6K1-mediated disassembly of mitochondrial URI/PP1gamma complexes activates a negative feedback program that counters S6K1 survival signaling. *Mol Cell.* 2007 Oct 12;28(1):28-40.

[49] Gu J., Li X., Liang Y., Qiao L., Ran D., Lu Y. et al., Upregulation of URI/RMP gene expression in cervical cancer by high-throughput tissue microarray analysis. *Int J Clin Exp Pathol.* 2013;6(4):669-77.

[50] Theurillat J. P., Metzler S. C., Henzi N., Djouder N., Helbling M., Zimmermann A. K. et al., URI is an oncogene amplified in ovarian cancer cells and is required for their survival. *Cancer Cell.* 2011 Mar 8; 19(3):317-32.

[51] Etemadmoghadam D., de Fazio A., Beroukhim R., Mermel C., George J., Getz G. et al., Integrated genome-wide DNA copy number and expression analysis identifies distinct mechanisms of primary chemoresistance in ovarian carcinomas. *Clin Cancer Res.* 2009 Feb 15; 15(4):1417-27.

[52] Hu J., Khanna V., Jones M. W., Surti U. Comparative study of primary and recurrent ovarian serous carcinomas: comparative genomic hybridization analysis with a potential application for prognosis. *Gynecol Oncol.* 2003 Jun;89(3):369-75.

[53] Lin L., Prescott M. S., Zhu Z., Singh P., Chun S. Y., Kuick R. D. et al., Identification and characterization of a 19q12 amplicon in esophageal adenocarcinomas reveals cyclin E as the best candidate gene for this amplicon. *Cancer Res.* 2000 Dec 15;60(24):7021-7.

[54] Schraml P., Bucher C., Bissig H., Nocito A., Haas P., Wilber K. et al., Cyclin E overexpression and amplification in human tumours. *J Pathol.* 2003 Jul;200(3):375-82.

[55] Leung S. Y., Ho C., Tu I. P., Li R., So S., Chu K. M. et al., Comprehensive analysis of 19q12 amplicon in human gastric cancers. *Mod Pathol.* 2006 Jun;19(6):854-63.

[56] Nakanishi M., Sakakura C., Fujita Y., Yasuoka R., Aragane H., Koide K. et al., Genomic alterations in primary gastric cancers analyzed by comparative genomic hybridization and clinicopathological factors. *Hepatogastroenterology*. 2000 May-Jun;47(33):658-62.

[57] Snijders A. M., Nowee M. E., Fridlyand J., Piek J. M., Dorsman J. C., Jain A. N. et al., Genome-wide-array-based comparative genomic hybridization reveals genetic homogeneity and frequent copy number increases encompassing CCNE1 in fallopian tube carcinoma. *Oncogene.* 2003 Jul 3;22(27):4281-6.

[58] Noske A., Henricksen L. A., LaFleur B., Zimmermann A. K., Tubbs A., Singh S. et al., Characterization of the 19q12 amplification including CCNE1 and URI in different epithelial ovarian cancer subtypes. *Exp Mol Pathol*. 2015 Feb;98(1):47-54.

[59] Bergmann A. Survival signaling goes BAD. *Dev Cell*. 2002 Nov;3(5): 607-8.

[60] Datta S. R., Ranger A. M., Lin M. Z., Sturgill J. F., Ma Y. C., Cowan C. W. et al., Survival factor-mediated BAD phosphorylation raises the mitochondrial threshold for apoptosis. *Dev Cell*. 2002 Nov;3(5):631-43.

[61] Harada H., Andersen J. S., Mann M., Terada N., Korsmeyer S. J., p70S6 kinase signals cell survival as well as growth, inactivating the pro-apoptotic molecule BAD. *Proc Natl Acad Sci USA*. 2001 Aug 14;98(17): 9666-70.

[62] Fan J. L., Zhang J., Dong L. W., Fu W. J., Du J., Shi H. G. et al., URI regulates tumorigenicity and chemotherapeutic resistance of multiple myeloma by modulating IL-6 transcription. *Cell Death Dis*. 2014;5: e1126.

# Bibliography

Abiotic stresses in crop plants LCCN 2014011777 Type of material Book Main title Abiotic stresses in crop plants / edited by Usha Chakraborty and Bishwanath Chakraborty. Published/Produced Boston, MA CABI 2014. Projected pub date 1411 Description pages cm ISBN 9781780643731 (hbk alk. paper) LC classification SB112.5 .A24 2014 Related names Chakraborty, Usha. Chakraborty, Bishwanath. C.A.B. International. Contents Heat shock proteins and molecular chaperones: role in regulation of cellular proteostasis and stress management -- Heat response, senescence and reproductive development in plants -- Ethylene, nitric oxide, and hemoglobins in plant tolerance to flooding -- Monitoring the activation of jasmonate biosynthesis genes for selection of chickpea hybrids tolerant to drought stress -- Genetic engineering crop plants to sustain drought tolerance -- Physiology and biochemistry of salt stress tolerance in plants -- Sugarcane (Saccharum sp.) salt tolerance at various developmental levels -- The impact of ozone pollution on plant defense metabolism: detrimental effects on yield and quality of agricultural crops -- Potentiality of ethylene in sulphur-mediated counteracting adverse effects of cadmium in plants -- Heavy metal and metalloid stress in plants: the genomics perspective -- Influence of arsenic and phosphate on the growth and metabolism of cultivated plants -- Plant responses to abiotic stresses in sustainable agriculture -- Interactive role of polyamines and reactive oxygen species in stress tolerance of plants -- Indirect and direct benefits of the use of Trichoderma harzianum strain T-22 in agronomic plants subjected to abiotic and biotic stresses -- Role of microorganisms in alleviation of abiotic stresses for sustainable agriculture. Subjects Crops--Effect of stress on. Crops--Physiology.

Archaea structure, habitats, and ecological significance LCCN 2010034002 Type of material Book Main title Archaea structure, habitats, and ecological significance / [edited by] Sakura Y. Kato. Published/Created Hauppauge, N.Y. Nova Science Publishers, c2010. Projected pub date 1103 Description p. cm. ISBN 9781617619328 (hardcover) LC classification QR82.A69 A738 2010 Related names Kato, Sakura Y. Contents The archaeal flagellum a novel prokaryotic motility structure and the role of posttranslational modifications in its assembly and function / Ken F. Jarrell -- Methanogenes and ammonia-oxidation archaeal communities in high temperature oil reservoirs / Hui Li ... [et al.] -- Diversity, function, and processing of archaeal non-coding RNAs / Kosuke Fujishima, Akio Kanai -- The archaea ribonuclease P molecular mechanisms of thermal adaptation / Francisco Miralles -- Mechanosensitive channels in archaea roles, lipid-protein interactions and future questions / Daniel Balleza -- The Nanoarchaeota physiology, genomics, and phylogeny / Ana Casanueva, Marla Tuffin, Don A. Cowan -- Chaperonin and prefoldin-two molecular chaperones that work cooperatively in archaea and eukaryotes / Tamotsu Zako ... [et al.] -- Primary production and carbon cycling in the deep sea archaeal carbon fixation and methane oxidation / Nianzhi Jiao ... [et al.]. Subjects Archaebacteria. Archaebacteria--Environmental aspects. Archaea. Biodegradation, Environmental. Ecosystem. Notes Includes bibliographical references and index.

Cellular trafficking of cell stress proteins in health and disease LCCN 2012951299 Type of material Book Main title Cellular trafficking of cell stress proteins in health and disease / Brian Henderson, A. Graham Pockley, editors. Published/Produced Dordrecht Springer, [2012] ©2012

Description xiii, 297 pages illustrations (some color) 24 cm ISBN 9789400747395 (hbk. alk. paper)) 940074739X (hbk. alk. paper) LC classification QP552.M64 C42 2012 Related names Henderson, Brian (Professor), editor of compilation. Pockley, A. Graham (Alan Graham), 1960- editor of compilation. Subjects Molecular chaperones. Eukaryotic cells. Prokaryotes. Proteins--Physiological transport. Heat shock proteins. Diseases. Molecular Chaperones. Heat-Shock Proteins. Protein Transport--physiology. Notes Includes bibliographical references and index. Series Heat shock proteins volume 6 Heat shock proteins (Series) v. 6.

Folding for the synapse LCCN 2010937760 Type of material Book Main title Folding for the synapse / Andreas Wyttenbach, Vincent O'Connor, editors. Published/Created New York Springer, c2011. Description viii, 318 p. ill. 24 cm. ISBN 9781441970602 (alk. paper) 1441970606 (alk. paper) 9781441970619 (e-book) 1441970614 (e-book) LC classification QH574 .F65 2011 Related names Wyttenbach, Andreas. O'Connor, Vincent, PhD. Subjects Cytology. Biochemistry. Neurosciences. Carrier proteins. Synapses. Physiology. Protein Folding. Molecular Chaperones. Synapses. Notes Includes bibliographical references and index.

Handbook of molecular chaperones roles, structures and mechanisms LCCN 2009037439 Type of material Book Main title Handbook of molecular chaperones roles, structures and mechanisms / Piero Durante and Leandro Colucci, editors. Published/Created New York Nova Biomedical Books, c2010. Description xx, 567 p. ill. (some col.) 27 cm. ISBN 9781608763665 (hardcover alk. paper) 1608763668 (hardcover alk. paper) LC classification QP552.M64 H36 2010 Related names Durante, Piero. Colucci, Leandro. Contents Molecular chaperones of the endoplasmic reticulum / Martin Schröder -- Type III secretion chaperones a molecular toolkit for all occasions / Matthew S. Francis -- Advances in heat shock proteomics towards a better understanding of the physiology and pathophysiology of molecular chaperones / Roberto Scatena, Patrizia Bottoni, Bruno Giardina -- System specific chaperones for membrane redox enzyme maturation in bacteria / Raymond J. Turner ... [et al.] -- Role of molecular biology in molecular imaging / Archana Mukherjee, Mathew L. Thakur -- Yeast protein aggregates, containing chaperones and glucose metabolism enzymes / O.V. Nevzglyadova ... [et al.] -- Structure and multiple functions of cyclophilin 40 a divergent loop cyclophilin / Thomas Ratajczak ... [et al.] -- Free-living protists as a model for studying heat shock proteins in the cell / Andrew V. Goodkov, Alexei O. Smurov, Yulia I. Podlipaeva -- Quality control in the secretory pathway / Joanna Jung ... [et al.] -- Structure, properties and multiple functions of human small heat shock protein HspB8 (Hsp22, H11 protein kinase or E2IG1) / Anton A. Shemetov ... [et al.] -- Small stress proteins and their therapeutic potential / Scott Houck, Joy G. Ghosh, John I. Clark -- The eukaryotic chaperonin CCT (TRiC) structure, mechanisms of action and substrate diversity / Julie Grantham -- The role of molecular chaperones in plant stress response / Youko Oono ... [et al.] -- Chaperone activity of intrinsically disordered proteins / Denes Kovacs, Peter Tompa -- Fluorescence study on aggregation of [alpha]-Crystallin and insulin / N.L. Vekshin -- The induction of hepatic heat shock protein 27 and 70 by an anti-hepatitis drug-bicyclol and its role in protection against liver injury / Geng-Tao Liu, Xiu-Qi Bao -- Plant heat shock proteins as molecular chaperones in normal and stress conditions / Bibliana Grigorova -- Different mechanisms responsible for stress resistance operate in the same insect order (diptera) / D.G. Garbuz ... [et al.] -- Bi-directional Hsp response in homeothermic livestock / Ariel Shabtay -- The chaperonin containing T-complex polypeptide do monomeric subunits have discrete individual functions? / Sandeep Kathju, Latha Satish -- Evidence for in vivo phosphorylation of Hsp26 from saccharomyces cerevisiae / D.N. Peixoto ... [et al.]. Subjects Molecular chaperones--Handbooks, manuals, etc. Molecular Chaperones. Notes Includes bibliographical references and index. Series Cell biology research progress series Cell biology research progress.

Inhibitors of molecular chaperones as therapeutic agents LCCN 2013444267 Type of material Book Main title Inhibitors of molecular chaperones as therapeutic agents / edited by Timothy Machajewski and Zhenhai Gao. Published/Created Cambridge Royal Soc Of Chemistry 2014 Published/Produced ©2014 Description xvi, 425 pages illustrations (some colored) 24 cm ISBN 1849736669 9781849736664 Related names Machajewski, Timothy D., editor of compilation. Gao, Zhenhai, editor of compilation. Rotella, David P., editor of compilation. Royal Society of Chemistry (Great Britain) Summary This book aims to provide a comprehensive examination of the field of molecular chaperone inhibition and its application to pharmaceutical research. With several small molecule inhibitors in oncology clinical development, there is clearly intense interest in the chaperones as a molecular target. Filling a significant gap in the market by providing a detailed comparison of discovery programs across the industry, this text will find broad interest among researchers in the field of molecular chaperone pharmaceutical research, oncology research, and medicinal chemistry. Arranged into three main sections the book covers structure and function, small molecule inhibitors and concludes with a section discussing clinical perspectives. With specific chapters covering the discovery of key molecules such as, BIIB028, STA-9090, Serenex Hsp90 inhibitor, NVP-AUY922 and NVP-HSP990, this comprehensive text will be an essential treatise for researchers working in academia and industry. Subjects Molecular chaperones. Chemical inhibitors. Series RSC drug discovery volume 37 RSC drug discovery series v.37.

Methods in protein biochemistry LCCN 2011029558 Type of material Book Main title Methods in protein biochemistry / edited by Harald Tschesche. Published/Created Berlin De Gruyter, c2012. Description xxiii, 352 p. ill. 25 cm. ISBN 9783110252330 (alk. paper) 3110252333 (alk. paper) 9783110252361 (e-ISBN) 3110252368 (e-ISBN) LC classification QP551 .M3883 2012 Related names Tschesche, Harald. Contents Three-phase partitioning -- Folding and degradation functions of molecular chaperones -- Membrane protein folding in detergents -- Glycoprotein folding quality control in the endoplasmic reticulum -- Protein import into the intermembrane space of mitochondria -- On-membrane identification of gel-resolved proteins by MALDI-MS -- Analysis of protein complexes using chemical cross-linking and mass spectrometry -- Single-crystal spectroscopy correlated with X-ray crystallography provides complementary perspectives on macromolecular function -- Wide-angle X-ray solution scattering (WAXS) -- Where purity matters recombinant versus synthetic peptides in beta amyloid formation -- Chemical modification of proteins in living cells -- Proteomics of human bronchoalveolar lavage fluid discovery of biomarkers of chronic obstructive pulmonary disease with difference gel electrophoresis (DIGE) and mass spectrometry -- Proteomic of Duchenne muscular dystrophy -- Target-oriented peptide arrays in a palliative approach to cystic fibrosis -- Probing protein dynamics in vivo using backbone cyclization bacterial acyl carrier protein as a case study -- The protein epitope mimetic approach to protein-protein interaction inhibitors -- The structural biology of alpha-1 antitrypsin deficiency and the serpinopathies. Subjects Proteins--Analysis. Proteomics--Methodology. Proteins--analysis. Proteins--chemistry. Proteins--isolation & purification. Proteomics--methods. Notes Includes bibliographical references and index.

Molecular chaperones methods and protocols LCCN 2011934798 Type of material Book Main title Molecular chaperones methods and protocols / edited by Stuart K. Calderwood, Thomas L. Prince. Published/Created New York Humana Press, c2011. Description xviii, 321 p. ill. (some col.) 26 cm. ISBN 9781617792946 (alk. paper) 1617792942 (alk. paper) 9781617792953 (e-ISBN) LC classification QP552.M64 M655 2011 Related names Calderwood, Stuart K. Prince, Thomas L. Contents Targeted deletion of Hsf1, 2, and 4 genes in mice -- The role of heat shock factors in stress-induced transcription -- Hsp90 and client protein maturation -- The role of p23, hop, immunophilins, and other co-chaperones in regulating hsp90 function -- Detecting HSP90 phosphorylation -- Role of molecular chaperones in biogenesis of the protein kinome -- Nucleotide exchange factors for hsp70 chaperones -- Reconstitution of CHIP E3 ubiquitin ligase activity -- Structure-functions of hspB1 (hsp27) -- Combined lentiviral and rnai technologies for the delivery and permanent silencing of the hsp25 gene -- Quantification of HSP27 and HSP70 molecular chaperone activities -- Measuring hsp72 (HSPA1A) by indirect sandwich ELISA -- Analysis of heat-shock protein localisation using flow cytometry -- Quantitation of heat-shock proteins in clinical samples using mass spectrometry -- Bioinformatic approach to identify chaperone pathway relationship from large-scale interaction networks -- Hsp70: Anti-apoptotic and tumorigenic protein -- Determination of cell survival or death -- Immunohistochemistry of human hsp60 in health and disease: From autoimmunity to cancer -- Preparation of a heat-shock protein 70-based vaccine from DC-tumor fusion cells -- Isolation of heat shock protein complexes -- Enhancing antigen cross-presentation and T-cell priming by complexing protein antigen to recombinant large heat-shock protein -- Investigating receptors for extracellular heat shock proteins -- Analysis of cellular migration using a two-chamber methodology. Subjects Molecular chaperones. Heat shock proteins. Notes Includes bibliographical references and index. Series Methods in molecular biology, 1940-6029 787 Springer protocols Springer protocols. Methods in molecular biology (Clifton, N.J.) 787.

Molecular chaperones LCCN 2012952678 Type of material Book Main title Molecular chaperones / Sophie Jackson, editor with contributions by A. Ahmad ... [et al.]. Published/Created Heidelberg New York Springer Verlag, c2013. Description ix, 272 p. ill. (some col.) 24 cm. ISBN 9783642345517 (alk. paper) 3642345514 (alk. paper) 9783642345524 (eBook) 3642345522 (eBook) LC classification QP552.M64 M63 2013 Related names Jackson, Sophie (Sophie E.) Ahmad, A. (Atta) Contents Assisting oxidative protein folding how do protein disulphide-isomerases couple conformational and chemical processes in protein folding? -- Peptide bond cis/trans isomerases a biocatalysis perspective of conformational dynamics in proteins -- Small heat-shock protein paramedics of the cell --

Allostery in the Hsp70 chaperone proteins -- Hsp90 structure and function -- Extracellular chaperones. Subjects Molecular chaperones. Protein folding. Molecular Chaperones. Protein Folding. Notes Includes bibliographical references and index. Additional formats Also issued online. Series Topics in current chemistry 0340-1022 328 Topics in current chemistry 328. 0340-1022.

Molecular chaperones of the endoplasmic reticulum LCCN 2010003161 Type of material Book Personal name Schröder, Martin 1954- Main title Molecular chaperones of the endoplasmic reticulum / Martin Schröder. Published/Created New York Nova Science Publishers, c2010. Description 126 p. ill. (some col.) 23 cm. ISBN 9781616681739 (softcover) 161668173X (softcover) LC classification QP552.M64 S37 2010 Contents Physicochemical principles governing protein folding -- The ER -- Molecular chaperones of the ER. Subjects Molecular chaperones. Endoplasmic reticulum. Endoplasmic Reticulum--physiology. Molecular Chaperones--physiology. Protein Folding. Notes Includes bibliographical references (p. [71]-119) and index. Series Protein Science and Engineering Protein science and engineering.

Protein aggregation LCCN 2010037892 Type of material Book Main title Protein aggregation / [edited by] Douglas A. Stein. Published/Created Hauppauge, N.Y. Nova Science Publishers, c2010. Projected pub date 1103 Description p. cm. ISBN 9781617618154 (hardcover) LC classification QP551 .P69545 2010 Related names Stein, Douglas A. Contents Changes in protein structure under the effects of cryopreservation and cryoprotective agents / N.G. Zemlianskykh -- Molecular chaperones and proteases as suppressors of protein aggregation in gram-negative bacteria / Joanna Skorko-Glonek ... [et al.] -- Native functions of amyloid / Reeba S. Jacob ... [et al.] -- Nucleation mechanisms and morphologies in insulin amyloid fibril formation / Vito Foderà ... [et al.] -- On the aggregation of albumin influences of the protein glycation / Philippe Rondeau ... [et al.] -- The role of conformational domain lability of fibrinogen molecules in processes of self-assembly of fibrin monomers and fibrinogen aggregation / M.A. Rosenfeld, V.B. Leonova, M.I. Biryukova -- Faced members of the family the synucleins / Andrei Surguchov -- Inclusion bodies a new concept of biocatalysts / Neus Ferrer-Miralles ... [et al.] -- Comparative study of bovine and ovine caseinate aggregation processes calcium-induced aggregation and acid aggregation / María Eugenia Hidalgo ... [et al.] -- Protein aggregation / Xiaoyu Lu, Yan Wei, Rongqiao He -- Yeast protein aggregates, containing chaperones, and glucose metabolism enzymes / O.V. Nevzglyadova ... [et al.] -- Folding and aggregation features of proteins / Oxana V. Galzitskaya, Sergiy O. Garbuzynskiy. Subjects Proteins--Denaturation. Protein folding. Protein Folding. Cell Aggregation. Protein Conformation. Notes Includes bibliographical references and index.

Protein chaperones and protection from neurodegenerative diseases LCCN 2010053393 Type of material Book Main title Protein chaperones and protection from neurodegenerative diseases / edited by Stephan Witt. Published/Created Hoboken John Wiley & Sons, c2011. Description xiv, 427 p., [4] p. of plates ill. 25 cm. ISBN 9780470569078 (cloth) LC classification QP552.M64 P76 2011 Related names Witt, Stephan. Subjects Molecular chaperones. Nervous system--Degeneration. Molecular Chaperones. Neurodegenerative Diseases--genetics. Notes Includes bibliographical references and index. Series Wiley series on protein and peptide science Wiley series in protein and peptide science.

The chaperonopathies diseases with defective molecular chaperones an introduction and guide to diseases in which chaperones play an etiologic-pathogenic role LCCN 2013931840 Type of material Book Personal name Macario, Alberto J. L., author. Main title The chaperonopathies diseases with defective molecular chaperones an introduction and guide to diseases in which chaperones play an etiologic-pathogenic role / Alberto J. L. Macario, Everly Conway de Macario, Francesco Cappello. Published/Produced Dordrecht New York Springer, [2013] ©2013 Description xvi, 116 pages illustrations (some color) 24 cm. ISBN

9789400746664 (alk. paper) 9400746660 (alk. paper) LC classification QP552.M64 M33 2013 Related names Conway de Macario, Everly, author. Cappello, Francesco, author. Summary This Brief provides a concise review of chaperonopathies, i.e., diseases in which molecular chaperones play an etiologic-pathogenic role. Introductory chapters deal with the chaperoning system and chaperoning teams and networks, HSP-chaperone subpopulations, the locations and functions of chaperones, and chaperone genes in humans. Other chapters present the chaperonopathies in general, including their molecular features and mechanistic classification into by defect, excess, or mistake. Subsequent chapters discuss the chaperonopathies in more detail, focusing on their distinctive characteristic. Subjects Molecular chaperones. Diseases--Causes and theories of causation. Medical microbiology. Molecular Chaperones--analysis. Disease--etiology. Protein Folding. Diseases--Causes and theories of causation. Medical microbiology. Molecular chaperones. Notes Includes bibliographical references. Series SpringerBriefs in biochemistry and molecular biology, 2211-9353 SpringerBriefs in biochemistry and molecular biology. 2211-9353.

Thermophilic microbes in environmental and industrial biotechnology biotechnology of thermophiles LCCN 2013931992 Type of material Book Main title Thermophilic microbes in environmental and industrial biotechnology biotechnology of thermophiles / Tulasi Satyanarayana, Jennifer Littlechild, Yutaka Kawarabayasi, editors. Edition Second edition. Published/Produced Dordrecht Springer Science+Business Media, [2013] ©2013 Description xx, 954 pages illustrations (some color), color maps 24 cm ISBN 9789400758988 (hd. bd. alk. paper) 9400758987 (hd. bd. alk. paper) LC classification TP248.27.M53 T48 2013 Related names Satyanarayana, T., editor. Littlechild, Jennifer, editor. Kawarabayasi, Yutaka, editor. Summary "The existence of life at high temperatures is quite fascinating. At elevated temperatures, only microorganisms are capable of growth and survival. Many thermophilic microbial genera have been isolated from man-made (washing machines, factory effluents, waste streams and acid mine effluents) and natural (volcanic areas, geothermal areas, terrestrial hot springs, submarine hydrothermal vents, geothermally heated oil reserves and oil wells, sun-heated litter and soils/sediments) thermal habitats throughout the world. Both culture-dependent and culture-independent approaches have been employed for understanding the diversity of microbes in hot environments. Interest in their diversity, ecology, and physiology has increased enormously during the past few decades as indicated by the deliberations in international conferences on extremophiles and thermophiles held every alternate year and papers published in journals such as Extremophiles. Thermophilic moulds and bacteria have been extensively studied in plant biomass bioconversion processes as sources of industrial enzymes and as gene donors. In the development of third generation biofuels such as bioethanol, thermophilic fungal and bacterial enzymes are of particular interest. The book is aimed at bringing together scattered up-to-date information on various aspects of thermophiles such as the diversity of thermophiles and viruses of thermophiles, their potential roles in pollution control and bioremediation, and composting."-- Back cover. Contents Thermophiles in the environment -- Diversity of hot environments and thermophilic microbes / Deepika Mehta, Tulasi Satyanarayana -- Exploring the ecology of thermophiles from Australia's Great Artesian Basin during the genomic era / Christopher D. Ogg, Mark D. Spanevello, Bharat K.C. Patel -- Hot environments from Antarctica: source of thermophiles and hyperthermophiles, with potential biotechnological applications / Patricio A. Flores, Maximiliano J. Amenábar, Jenny M. Blamey -- Bacterial and biochemical properties of newly invented aerobic, high-temperature compost / Takahiro Yoshii, Toshiyuki Moriya, Tairo Oshima -- Role of thermophilic microflora in composting / Seema Rawat, Bhavdish Narain Johri -- Metal bioremediation by thermophilic microorganisms / Pinaki Sar, Sufia K. Kazy, Dhiraj Paul, Angana Sarkar -- CO-oxidizing anaerobic thermophilic prokaryotes / T. Sokolova, A. Lebedinsky --

Biomineralization in geothermal environments / Katsumi Doi, Yasuhiro Fujino -- Phylogeny and biological features of thermophiles / Takashi Itoh, Takao Iino -- Biology, biodiversity and application of thermophilic viruses / Kristine Uldahl, Xu Peng -- Genomics, metagenomics and biotechnology -- Genomics of thermophilic bacteria and archaea / Takaaki Sato, Haruyuki Atomi -- Comparative genomics of thermophilic bacteria and archaea / Satoshi Akanuma, Shin-ichi Yokobori, Akihiko Yamagishi -- Host-vector systems in thermophiles / Takahiro Inoue, Yoshihiko Sako -- Molecular chaperones in thermophilic eubacteria and archaea / Muhamad Sahlan, Masafumi Yohda -- Heterologous production of thermostable proteins and enzymes / Haruhiko Sakuraba, Toshihisa Ohshima -- Discovery of thermostable enzymes from hot environmental samples by metagenomic approaches / Norio Kurosawa -- DNA polymerases and DNA ligases / Sonoko Ishino, Yoshizumi Ishino -- Molecular diversity and biotechnological relevance of thermophilic actinobacteria / Satya P. Singh, Rushit J. Shukla, Bhavtosh A. Kikani -- Mechanisms of thermal stability adopted by thermophilic proteins and their use in white biotechnology / Jennifer Littlechild, Halina Novak, Paul James, Christopher Sayer -- Starch-hydrolyzing enzymes from thermophiles / Skander Elleuche, Garabed Antranikian -- Thermostable archaeal and bacterial pullulanases and amylopullulanases / M. Nisha, Tulasi Satyanarayana -- Sugar metabolic enzymes / Kazuaki Yoshimune, Yutaka Kawarabayasi -- Restriction enzymes from thermophiles / Prince Sharma, Ravinder Kumar, Neena Capalash -- Microbial chitinases: natural sources, mutagenesis, and directed evolution to obtain thermophilic counterparts / Pullabhotla Venkata Subba Rama Narsimha Sarma, Jogi Madhu Prakash, Subha Narayan Das -- Phytases and phosphatases of thermophilic microbes: production, characteristics and multifarious biotechnological applications / Bijender Singh, Tulasi Satyanarayana -- Pectinases of thermophilic microbes / Saurabh Sudha Dhiman, Ritu Mahajan, and Jitender Sharma -- Developments in thermostable gellan lyase / Margarita Kambourova, Anna Derekova --

The lignocellulolytic system of thermophilic fungi and actinomycetes: structure, regulation, and biotechnological applications / Marcio José Poças-Fonseca, Robson Willian de Melo Matos, and Thiago Machado Mello-de-Sousa -- Cellulases of thermophilic microbes / Linga Venkateswar Rao, Anuj K. Chandel, G. Chandrasekhar, A. Vimala Rodhe, J. Sridevi -- Xylanases from thermophilic fungi: classification, structure, and case study of Melanocarpus albomyces / Saroj Mishra, Vikram Sahai, Virendra Swaroop Bisaria, Ranjita Biswas, Gupteshwar Gupta, Swati Nakra -- Thermostable bacterial xylanases / Vikash Kumar Digvijay Verma, A. Archana, Tulasi Satyanarayana -- Thermostable proteases / Rajeshwari Sinha, Sunil K. Khare -- Microbial keratinases: diversity and applications / Rani Gupta, Ekta Tiwary, Richa Sharma, Rinky Rajput, Neha Nair -- Biocatalysis through thermostable lipases: adding flavor to chemistry / Rohit Sharma, Vishal Thakur, Monika Sharma, Nils-Kåre Birkeland. Subjects Thermophilic microorganisms. Microbial biotechnology. Hot Temperature. Microbiological Phenomena. Adaptation, Physiological. Archaea--enzymology. Archaea--genetics. Bacteria--enzymology. Bacteria--genetics. Biotechnology. Microbial biotechnology. Thermophilic microorganisms. Notes Includes bibliographical references and index.

# INDEX

## A



## B

## D



## E

## F



## G



## H

## M



## N



## O

## P

## Q



## R



## S

## W



## Y